To [illegible]
with gratitude for
traveling so many miles
on those roads leading
to new frontiers.

Dr. Shields

AUTHOR

Air Emissions, Baselines, and Environmental Auditing

Edited by Jacqueline Shields

VAN NOSTRAND REINHOLD
New York

Library of Congress Catalog Card Number 93-6935
ISBN 0-442-01092-3

I(T)P Van Nostrand Reinhold is an International Thomson Publishing Company.
ITP logo is a trademark under license.

Printed in the United States of America.

Van Nostrand Reinhold
115 Fifth Avenue
New York, New York 10003

International Thomson Publishing
Berkshire House, 168-173
High Holborn
London WC1V7AA
England

Thomas Nelson Australia
102 Dodds Street
South Melbourne 3205
Victoria, Australia

Nelson Canada
1120 Birchmount Road
Scarborough, Ontario
M1K 5G4, Canada

International Thomson Publishing GmbH
Königswinterer Str. 518
5300 Bonn 3
Germany

International Thomson Publishing Asia
38 Kim Tian Rd., #0105
Kim Tian Plaza
Singapore

International Thomson Publishing Japan
Kyowa Building, 3F
2-2-1 Hirakawacho
Chiyada-Ku, Tokyo 102
Japan

16 15 14 13 12 11 10 9 8 7 6 5 4 3 2 1

Library of Congress Cataloging-in-Publication Data

Air emissions, baselines, and environmental auditing / edited by J. Shields
p. cm.
Includes bibliographical references and index.
ISBN 0-442-01092-3
1. Air quality management. 2. Environmental auditing. 3. Environmental auditing—Case studies. I. Shields, J. (Jacqueline)
TD883.1.A362 1993
363.73'9263—dc20 93-6935
CIP

Contents

Preface vii

Part I Introduction and Legal Requirements 1

Chapter 1 The History of Environmental Auditing 3
by Frank J. Priznar
THE RENAISSANCE PERIOD 3
A PERIOD OF DIVERGENCE 5
THE AGE OF ENLIGHTENED THINKING 7
THE FUTURE 8

Chapter 2 Environmental Legislation and Guidelines 10
by William P. Gulledge
INTRODUCTION 10
ENVIRONMENTAL AUDITING THEMES OF LEGISLATION AND POLICIES 11
IMPACT OF LEGISLATION AND GUIDELINES ON ENVIRONMENTAL AUDITING 15

Part II Principles and Practices 19

Chapter 3 Auditing Techniques 21
by Lisa E. Lowe
INTRODUCTION 21
AUDIT GOALS AND OBJECTIVES 21
PLANNING THE AUDIT 24
PREAUDIT ACTIVITIES 31
SITE VISIT 34
DATA EVALUATION 35
AUDIT REPORT 36
ACTION PLANS 37
EVALUATION OF AN AUDIT PROGRAM 38
CONCLUSIONS 39

Chapter 4 Quality Assurance for Environmental Auditing Programs 41
by Michael D. Henke and Jill Barson Gilbert
INTRODUCTION 41
QUALITY ASSURANCE OF ENVIRONMENTAL AUDIT PROGRAMS 42
QUALITY ASSURANCE MECHANISMS 43
CONDUCTING QUALITY ASSURANCE REVIEWS 46

COMPONENTS OF ENVIRONMENTAL AUDIT PROGRAMS 48
SUMMARY 53

Chapter 5 Employment of an Environmental Auditor 54
by Jacqueline Shields and Jill Barson Gilbert
PURPOSE OF THE AUDIT 54
HIRING OPTIONS 55
ESTABLISHING BOUNDARIES 57
CRITERIA FOR HIRING THE AUDITOR 57
CRITERIA CATEGORIES 58
FINAL SELECTION OF THE AUDITOR 61

Part III Specific Situations 63

Chapter 6 Air Quality Data Sources for Audits 65
by William P. Gulledge
INTRODUCTION 65
AIR QUALITY DATA SOURCES 65
TYPES OF DATA TO BE AUDITED 69
Appendix 6-A—Air Pollution Audit Materials 71

Chapter 7 Monitoring 83
by William P. Gulledge
MONITORING 83
AUDITING OF AIR MONITORING RESULTS 85

Chapter 8 Baselines 87
by Jacqueline Shields
TYPES OF BASELINES 87
ESTABLISHING BASELINES 88
BASELINES IN THE 1990 CAA 90
SUMMARY OF BASELINE REQUIREMENTS 92
ADJUSTMENTS IN AUDITING UNDER THE 1990 CAA 101
Appendix 8-A—Baseline Requirements 103

Chapter 9 Air Permit Audits 115
by Jacqueline Shields and Paul D. Rosenstock
GENERAL REQUIREMENTS FOR AIR PERMIT AUDITING 115
ADMINISTRATIVE AUDIT 116
COMPLIANCE AUDIT 122
AIR PERMITTING UNDER THE 1990 CAA AMENDMENTS 125

Chapter 10 Air Emissions, Environmental Auditing, and Waste Minimization 130
by Paul D. Rosenstock and Sam N. Popowcer
THE REGULATORY ENVIRONMENT 130
THE 1990 CLEAN AIR ACT AMENDMENTS 133
THE ENVIRONMENTAL AUDIT 137
WASTE MINIMIZATION 145

Chapter 11 Property Transactions 149
by Jill Barson Gilbert and Douglas J. Hileman
INTRODUCTION 149
ENVIRONMENTAL DUE DILIGENCE 150
REGULATORY FRAMEWORK 151
ENVIRONMENTAL DUE DILIGENCE 159
THE PROPERTY LIFE CYCLE 168
ENVIRONMENTAL ISSUES AND PROPERTY TRANSACTIONS 175
SUMMARY 183

Chapter 12 Environmental Audits at Refineries and Petrochemical Facilities 185
by David A. Dougall and Jill Barson Gilbert
INTRODUCTION 185
OBJECTIVES 186
APPROACH 187
REPORT 189
ENVIRONMENTAL ISSUES 192
SUMMARY 205

Chapter 13 An Audit Approach to Source Reduction of Air Toxics and Ozone-Depleting Chemicals 207
by Pamela Beilke and Linda Micale
INTRODUCTION 207
OVERVIEW OF POLLUTION PREVENTION PROGRAMS 208
THE FEDERAL PROGRAM—TOXIC CHEMICAL SOURCE REDUCTION AND RECYCLING 209
A CLOSER LOOK AT AIR TOXICS AND OZONE-DEPLETING CHEMICALS 211
AN AUDIT APPROACH TO SOURCE REDUCTION OF AIR TOXICS AND OZONE-DEPLETING CHEMICALS 213

Chapter 14 The Facility Environmental Health Audit 218
by Jacqueline Shields
INTRODUCTION 218
LEGISLATION 220
STRUCTURE OF THE ENVIRONMENTAL HEALTH AUDIT 229

CONDUCTING THE ENVIRONMENTAL HEALTH AUDIT 237
SUMMARY 241

Part IV Case Studies 243

Case Study #1—Reductions in Air Emissions of Ozone-Depleting Chemicals 245
by Pamela J. Beilke

Case Study #2—Phased Chemical Usage Study and Authorization Review—Electronics Manufacturer 247
by Linda Micale

Case Study #3—Hazardous Materials System and Waste Minimization Project—Electronics Manufacturing Facility 249
by Linda Micale

Case Study #4—Volatile Organic Chemical Inventory—Nuclear Generating Station (NGS) 251
by Linda Micale

Case Study #5—Waste Minimization 252
by Paul R. Rosenstock and Sam N. Popowcer

Case Study #6—Refinery Environmental Audit 254
by Paul R. Rosenstock and Sam N. Popowcer

Case Study #7—Refinery Environmental Audit 257
by Jill Barson Gilbert
SITE HISTORY 257
ENVIRONMENTAL ISSUES 257

Case Study #8—Purchase of Fertilizer Plant 260
by Jill Barson Gilbert
SITE HISTORY 260
ENVIRONMENTAL ISSUES 260
CONCERNS 262

Case Study #9—A Regional Ambient Air Quality Audit 263
by Jacqueline Shields

Contributors 266

Index 273

Preface

Environmental auditing is rapidly becoming a mature environmental field, as is reflected in the emergence of a common language and standardization of practices. However, some variations in terminology and practice still remain and are reflected in the slight variations in each author's language and style of presentation in each chapter. These differences have been retained to accurately reflect the current terminology and practices in environmental auditing. This caveat is provided so that the reader will be aware of this transitional period and the exciting changes that are occurring in the field.

Within the next few years, major changes in environmental auditing will occur as the result of the passage of the Clean Air Act Amendments in 1990. Not only are compliance audits by business and industry required under the Act, but there are extensive requirements for assessment audits to be conducted by governmental agencies. The Act is divided into 11 major sections, or Titles, covering everything from the pollutants that will be regulated to the establishment of new air quality control programs. This will require many changes, not only in air quality management, but also in the life-styles of individual citizens. The full extent of these changes will not be known until the final regulations are promulgated approximately 4 years from now.

In an effort to prepare to meet these new, far-reaching regulations, many companies have already initiated audits of air emissions regulated under the Act. This book will assist companies in these efforts by addressing both the assessment and compliance auditing of air emissions. The book is divided into four major sections: legal requirements, principles and practices, specific situations, and case studies. A balance of theory and practice is presented in each chapter, with discussions of problems, caveats, and tips on things that work in the real world. Each chapter has been written by practitioners in the field of environmental auditing, with emphasis on auditing air emissions. This book is the first to examine the environmental auditing of air emissions in the context of the Clean Air Act Amendments of 1990. Hopefully, it will serve as the baseline from which future audits of air emissions will be conducted.

Jacqueline Shields
San Antonio, Texas

I

Introduction and Legal Requirements

1

The History of Environmental Auditing

Frank J. Priznar

The history of environmental auditing is perhaps best thought of in terms of three general periods. The first, from about 1977 to about 1983, can be considered a Renaissance period. During this time, the term "environmental auditing" was coined, widely used, and broadly understood among practitioners. The second period, one of divergence, was characterized by multidimensional growth. It lasted until about 1989, although remnants of it remain. In this period, environmental auditing grew to mean different things to different people, creating a mass of confusion among both practitioners and the entities being audited. Finally, a period of enlightened thinking emerged a few years ago when environmental auditing gained recognition as a profession and a supporting infrastructure started to take shape and be recognized.

Admittedly, these dates are imprecise, and there is some overlap in the events that characterize the periods; however, many experts who have been involved in environmental auditing for many years would agree with the segmentation. Because of this, it is convenient to discuss the history of environmental auditing in terms of these periods and to focus on the factors that drove them and the characteristics they displayed.

THE RENAISSANCE PERIOD

There are at least three documented accounts that describe the origin of environmental auditing (Palmisano 1989; Rhodes 1989; Cahill and Kane 1989). Although the views are somewhat different, many of the facts cited are common, leading many to accept that formal environmental auditing was initiated by Security and Exchange Commission (SEC) enforcement actions against three large companies—the first against Allied Chemical Corporation in 1977, the second against United States Steel in 1979, and the third against Occidental Petroleum in 1980. In short, the SEC required that each of these companies more accurately determine their financial liability from environmental issues through a corporate-wide environmental auditing program. They responded at the time and, to their credit, each of these firms now has a first-class environmental auditing program in place that is often cited as a model for others.

The details of these SEC actions are no longer as important as they were at the time because industry apparently got the message about disclosing financial liability due to environmental issues. This is evidenced by the fact that there have been many environmental audits completed since these benchmark cases, but very few were driven by SEC actions. This will probably remain true as long as the SEC believes that corporations are properly accounting for environmental issues in their financial statements.

During this period, environmental auditing usually meant environmental compliance auditing. More specifically, audits were done to ascertain whether a facility or operation was or was not in compliance with applicable environmental laws. This helped corporations understand where there was environmental risk for fines or penalties due to noncompliance (a key issue) and, more importantly, where financial resources would be needed to meet environmental standards.

Around 1979, the Environmental Protection Agency (EPA) learned about environmental auditing through a contracted study on innovative enforcement approaches and began its own research into its value as a compliance assurance mechanism. Through its contacted studies and its own published reports, the EPA, perhaps intentionally, promoted environmental auditing by making it a visible and debatable topic. Shortly, environmental auditing became a buzz word inside and outside of the EPA. Consultants began to sell the benefits of environmental auditing to numerous industries who were bewildered by the explosion in complex environmental regulations and an apparent need for third-party verification of compliance status. Other industries emerged voluntarily as practitioners, such as Olin, Amoco, and General Motors, and the EPA continued to make it a visible subject, although it never seemed able to clearly define its own role substantially.

In 1981, some of the most interested industrial practitioners began meeting on a periodic basis to compare their approaches and openly discuss their experiences in achieving perceived goals of a successful environmental auditing program. This group formed the nucleus of what eventually became the Environmental Auditing Roundtable (EAR), which still meets three times per year in different locations around the United States. As a forum for advancing the practice of environmental auditing, the EAR served as the earliest and most important catalyst.

Although there was recognition and some growth in environmental auditing during this period, it was still not widely used in the mainstream of corporate environmental management tools and techniques. One reason for this is that most industries were too busy reacting to environmental requirements to have time to be proactive, which is a fundamental requirement of environmental auditing. Another reason is perhaps that it was being practiced, but it was called something else. One source (Palmisano 1989) states that in 1978 there was only one published report, study, book, or article on environmental auditing, compared to 29 in 1983. Five of these were grouped in a special session of a 76th meeting of the then-called Air Pollution Control Association (Cogen 1983), now known as the Air and Waste Management Association. Clearly, by the end of this period, the word was getting out.

Renaissance seems an appropriate name for this period because the individual environmental auditing leaders of the period had broad intellectual and cultural interests that encompassed the full spectrum of available and relevant knowledge. This included the principles of financial auditing (an important reference for environmental auditing), business management, and environmental requirements and issues, among others. Because environmental requirements and issues became so complicated after this period, it was incredibly difficult, if at all possible, for individuals to ever achieve this complete level of understanding in environmental auditing again. Thus the Renaissance period ended when complexities caused divergence, the underlying characteristics of the next period.

A PERIOD OF DIVERGENCE

After 1983, a large number of parties entered the environmental auditing arena with differing objectives that broke the earlier sense of unanimity about what it was and why it was done. Some of the factors that played heavily were:

- Industry's fear that the EPA would launch a mandatory environmental auditing program with no controls on disclosure of findings and other threatening but unknown characteristics;
- The EPA's slowness in taking a meaningful and clearly understood stand; and
- The growing need and desire to move out on environmental auditing by everyone.

In essence, everyone thought it was a good idea, and many companies were willing to invest time and money into it, but there was no way to clearly read the playing field. Industry, for example, did not want to suffer negative consequences from the EPA for doing what it thought was a fundamentally good idea. There should have been rewards or positive recognition, but these were not guaranteed.

These clashing forces, as strong as they were to those who were watching, were invisible to many who just "discovered" environmental auditing. Driven mainly by concerns about financial liability from hazardous waste disposal practices under the Comprehensive Environmental Response, Compensation, and Liability Act (CERCLA) of 1980, the Superfund Amendments and Reauthorization Act (SARA) of 1986, the Resource Conservation and Recovery Act (RCRA) in 1976, and emerging state-level laws, property owners, sellers, buyers, and leaders learned that there was value in having an assessment of these liabilities. The consulting community responded enthusiastically to the need for these assessments and in phases (e.g., Phase I assessment) provided thousands upon thousands of these. In many, if not most, cases the assessments were promoted as environmental audits, although they were very different from what veterans of the renaissance period had in mind.

Environmental liability site assessments (ELSAs), also known by perhaps at least a dozen other names, became the predominant environmental "audit" types of the period. Compliance audits were still completed, but their numbers were small in comparison. Unfortunately, when a firm wished to contract out for an audit, it was not always clear what the proposals were offering. In these cases, the low bidder was probably offering to complete an ELSA as opposed to a justifiably more costly compliance audit. Beyond this confusion, there were no standards for any of these audit types, making comparisons even more difficult.

One positive thing the ELSAs did for environmental auditing was to promote the term and the overall concept more broadly. By the end of this period, anyone in the environmental business who had not heard of environmental auditing, of one kind or another, must have been working in total isolation. There were now hundreds of articles written, books published, courses offered throughout the United States on how to conduct an audit, specialty conferences held, and newsletters published.

Somewhere near the center of this period, in 1986, pivotal events occurred in environmental auditing. The EPA published its environmental auditing policy statement in the Federal Register (51 Fed. Reg 25004). In essence, the statement encouraged environmental auditing among regulated industry. In a separate but much less publicized action later that year, the EPA decided to issue internal policy on when to include environmental auditing in enforcement settlements. With this approach, the EPA offered a two-pronged approach to compliance. One hand encouraged industry to comply, while its other hand was clenched in a fist ready to punish those who did not. Not everyone got the message until too late, as the EPA soon had dozens of enforcement settlements in many media programs that included mandatory environmental audits.

Other key events of 1986 included the formation of the Institute for Environmental Auditing (IEA) in Washington, D.C. and the passage in California of SB 1875, which resulted in a statewide voluntary environmental assessor program. The IEA was formed by dissatisfied EAR members who wished to create a professional association of environmental auditors with standards and credentialing programs that the EAR leadership was not ready to accept. For other reasons, another group, the Environmental Auditing Forum, split off from the EAR at about the same time, with California as a geographic base of membership. The fractioning of serious environmental auditing professionals and experts contributed, unfortunately, to a loss of momentum among the Renaissance devotees that is only now regaining strength. The key issue was that during this period there was no single body of respected environmental auditors to speak for the profession. In fact, other professional associations were formed for specific geographic territories of environmental auditors or for specific audit types, such as the ELSAs.

The California Bill, on the other hand, demonstrated that there was a great deal of interest in auditing as a profession, although most registrants still thought of assessments as audits. Within a few years of its creation, several thousand individuals paid to be registered in California as an assessor, even though it was not a requirement anywhere for anything. This public demonstration of desire for recognition was just further evidence that environmental auditing was a valued service and that its practitioners sought guidance and recognition.

In summary, this period saw environmental auditing become a common practice, although there was not unanimity on details. The number of practitioners grew from perhaps a hundred or so to thousands, corporations grew somewhat wary of the EPA, and separate groups of professional environmental auditors formed. By the close of this period, it was clear that some order needed to be made of environmental auditing in the United States, as other countries were asking for advice in developing their own programs.

THE AGE OF ENLIGHTENED THINKING

In 1989, the International Chamber of Commerce hosted a symposium on environmental auditing in France and subsequently published its own policy statement. Although it was essentially compatible with the EPA's statement, the goals of environmental auditing suddenly took on global significance and controversial issues that seemed monumental in the United States became much less important. Suddenly it seemed there was additional support for environmental auditing, and some logical steps, long overdue, were taken.

Some clear trends were emerging in environmental auditing (Priznar 1990) that were often the focus of discussion. These included standards, the use of automated information systems, auditor qualifications, and others. At least now it was possible to agree on the topics that were to be addressed, even though there was no general agreement yet on the outcomes. Additionally, there were clearer categories of audit types that could be characterized and that gave some structure to standard setting and qualifications of practitioners (Priznar 1991).

Sensing the value that environmental auditing could bring to more formal regulated environmental management programs, politicians began proposing laws that included environmental auditing. These included new versions of the Clean Water Act and the Clean Air Act. In fact, at this point, few of these have passed, but their appearance at the state and federal level every year suggest that the promotion, it not control, of environmental auditing by regulation is just a matter of time in the United States. In Europe, it seems to be taking a similar path through the eco-auditing scheme.

Recently, the Department of Justice (DOJ) and again the EPA stated or restated a policy on environmental auditing. The DOJ statement, issued in July 1991, was titled "Factors in Decisions on Criminal Prosecutions for Environmental Violations in the Context of Significant Voluntary Compliance of Disclosure Efforts by the Violator." It was supposed to offer some assurance to industry that its voluntary efforts would be favorably taken into account if there were environmental enforcement actions taken against it. For the EPA, which published a policy statement on December 12, 1991, in the Federal Register, a similar approach was offered. In both cases, the federal government was essentially saying, "We appreciate your efforts to comply with environmental laws, and will take as evidence of your sincerity to comply, the existence of a formal environmental auditing program." At this time, it appears that everyone in the government, industry, and the practitioners are taking steps to finally reach agreement on environmental auditing's value and to work towards making it a recognized profession.

THE FUTURE

The future is not really part of history; however, it's hard to resist making some predictions in a chapter like this. If we can learn from history, then the following are likely to occur within the next few years. First, a consolidated and international organization of environmental auditors will emerge as an accepted body of experts whose opinion will be sought on proposed legislation and standards that will affect them. This organization may use some of the international standards as a rallying point. Second, the EPA will ultimately endorse qualifications of environmental auditors to fill gaps left by legislation; otherwise it will be left up to the individual states. International issues, such as trade, global warming, and sustainable development, will have a more profound impact on the need for environmental auditors than most would guess at this time. This is because of the transboundary characteristics of environmental issues such as competitive strategy and air pollution. Finally, a lot of what we now consider to be environmental auditing will be completed by automated information systems, as auditing is essentially a process of verification.

References

Palmisano, J. 1989. Environmental auditing: past, present, and future. *Environmental Auditor.* Vol. 1, No. 1.

Rhodes, R. L. 1989. The U.S. experience as seen by industry. *Environmental Auditor.* Vol. 1, No. 3.

Cahill, L. B., and R. W. Kane, Ed. 1991. *Environmental Audits.* 6th ed. Rockville, Md. Government Institutes.

Cogen, Richard, Chair. 1983. Environmental Auditing. Paper read at Session 19 of the 76th Annual Meeting of the Air Pollution Control Association, 19–24 June 1983, in Atlanta, Ga.

Priznar, F. J. 1990. Trends in Environmental Auditing. *Environmental Law Reporter.* 20. ELR 10179, Washington, D.C.

Priznar, F. J. 1991. Audit types and their appropriate use. Paper read at UNEP/IEO Institutional Expert Seminar, 8 April, at Trolleholm Castle. Available from the Department of Industrial Environmental Economics, Lund University, Lund, Sweden.

2

Environmental Legislation and Guidelines

William P. Gulledge

INTRODUCTION

Most of the important federal environmental statutes have been enacted within the past 20 years. Prior to 1970, the year the U.S. Environmental Protection Agency (EPA) was created and the first significant environmental statute was passed, most environmental concerns were mildly regulated at the state level or through independent commissions within the federal government. In 1970, on the public initiative that culminated with the first Earth Day, Congress began enacting a series of laws that would affect businesses and life-styles over the next 20 years. This effort continues with the passage of the 1990 Clean Air Act (CAA) Amendments and consideration of future, stronger pollution prevention requirements.

The most important federal statutes that include some aspect of environmental auditing or predecessor studies include the following legislation and significant amendments:

- CAA Amendments in 1977, with Amendments in 1990;
- Resource Conservation and Recovery Act (RCRA) in 1976, with amendments in 1984;
- Comprehensive Environmental Response Compensation and Liability Act (CERCLA) in 1980;
- Superfund Amendments Reauthorization Act (SARA) in 1986;
- Federal Water Pollution Control Act (FWPCA) Amendments in 1972, with amendments in 1977, 1980, and 1987;
- Toxic Substances Control Act (TSCA) in 1976;
- National Environmental Policy Act (NEPA) in 1970; and
- Safe Drinking Water Act (SDWA) Amendments in 1986.

The CAA establishes the base requirements for compliance with air quality standards. Reductions in pollutants emitted into the air are achieved through a combination of air quality standards, stationary source emission limits, new source emission reductions, and mobile source limits. Traditional approaches to reducing air emissions have focused on technology-based source controls. The current CAA provides for a combination of source controls and market incentives to reduce air pollution.

Several of the laws listed above plus additional policies and guidelines have helped frame current environmental auditing practices. The most significant environmental auditing-related provisions include:

- 1990 CAA Amendments, Section 504, on compliance certification and reporting requirements for permits;
- SARA Title III emissions reporting requirements;
- TSCA Section 8 substantial risk information and disclosure of health and safety studies;
- Environmental auditing policy statement (1986); and
- Department of Justice (DOJ) guidance regarding environmental audits (1991).

An important distinction should be made between an "environmental audit" and an "environmental assessment." The term "environmental audit" is used by most environmental professionals to denote a process for independently determining an entity's compliance with environmental policies or regulations. An "environmental assessment" is normally not an audit at all, but an assessment of a site's or an entity's environmental condition or risk. An assessment may or may not include a compliance review. The user and producer of an assessment or audit should delineate the scope of either investigation.

ENVIRONMENTAL AUDITING THEMES OF LEGISLATION AND POLICIES

Technology-based and Performance-based Approaches

Developing auditing programs from environmental legislation or policies can be confusing. It is important to understand the difference between technology-based regulation and performance-based regulation. Technology-based standards are best represented by emission limits or design standards. Many in the environmental management profession have designated this the "command and control" approach.

Most environmental laws originated with a technology-based approach for environmental regulation. Little judgement is left to the regulated community on how to achieve these standards. If you don't achieve the emissions limitation number or comply with the design standard, you are not in compliance. Many regulatory standards developed in the 1970s and early 1980s also applied to industrial categories or types of industries. For instance, wastewater effluent guidelines and treatment standards were developed for separate industrial categories—organic chemicals, inorganic chemicals, petroleum refining, pulp and paper manufacturing, electric utilities, and so on.

A performance-based environmental regulation is generally designed to offer the regulated entity more flexibility in achieving compliance. A cleanup goal or a water quality–based standard are examples of a performance-based standard. Assuming no technology-limiting permit conditions, the affected party would be free to choose any approach of meeting this cleanup goal as long as human health and the environment were being adequately protected. In practice, performance-based pollution control approaches have been difficult to implement in the United States. Congress, environmental groups, and some industries have resisted the inherent flexibility of this approach. Early performance-based efforts were seen as reducing environmental control and increasing risk to human health or the environment.

The CAA, the FWPCA, and even the RCRA represent the early focus of technology-based standards and guidelines for pollution control. Under the CAA, technology-based, numerical emission limitations designed to achieve National Ambient Air Quality Standards (NAAQS) are written into permits. Auditing for compliance under this approach appears straight forward—you either comply with emission limits or you don't. However, this determination is much harder to make in the real world. Frequently, permit limitations cannot be achieved all the time. Exceedances and process upsets do occur. Emissions control technology is not perfect, and neither is the scientific basis used to set air quality standards and limits.

As an example, one issue that continues to be debated is what averaging time or time period should be used to determine compliance. Many NAAQSs were created with an averaging time automatically assumed—anywhere from 1 to 24 hours. The standard and monitoring requirements for permit or ambient air monitoring conditions may contain different time periods.

Anyone who completes compliance audits realizes that no emissions control technology is capable of controlling emissions 100% of the time, but measuring compliance with a technology-based standard is generally easier than determining compliance with a performance-based environmental standard. The environmental auditor, using a compliance checklist, can easily determine whether required monitoring is being conducted at a facility or whether pollution control equipment is generally capable of achieving emission limits.

The environmental auditor faces a tougher challenge in attempting to audit a corporate requirement to maintain "good housekeeping" at a facility or to conform to a corporate policy that calls for going beyond compliance with environmental regulations. There are no quantitative parameters to audit against. Auditing criteria for performance-based parameters must be developed before an effective audit can be conducted.

Companies have begun to audit performance-based parameters. Criteria are being established to assess pollution prevention effectiveness or beyond-compliance environmental management. The Institute for Environmental Auditing recently adopted a standard for auditing environmental management programs (included at the conclusion to this chapter). This activity is especially important under the CAA Amendments of 1990.

The 1990 CAA is the most confusing, complex environmental statute ever passed by Congress. Over 30 assessment or audit types are included in the CAA. Most of these provisions focus on environmental health assessments.

It has been generally referred to as a performance-based environmental statute. It includes extensive use of emissions trading and flexible permit procedures to achieve reductions in air pollution. Section 504 requires compliance certification and reporting requirements for permits. Many companies are now focusing on establishing auditing criteria to be used in the permit applications.

Environmental Auditing Policy Statement

Although now somewhat dated, the environmental auditing policy statement issued by the EPA in 1986 has strongly influenced the field of environmental auditing. This statement grew out of several environmental auditing approaches being debated within the EPA during the early 1980s. The first approach investigated by the EPA was to use third-party auditors to determine compliance with National Pollutant Discharge Elimination System (NPDES) permit requirements. The Council on Environmental Quality (CEQ) originally proposed this approach in the late 1970s.

Under the CEQ proposal, certified environmental auditors would be retained by the discharger to determine, by reviewing the monitoring and compliance programs, whether permit requirements were being met. This approach was not well received by the regulators or the regulated dischargers. Perceived high cost to both parties was the major reason this approach was rejected. Concerns over liability and disclosure of enforcement-related information also contributed to the proposal's failure.

In 1981, the EPA began promoting incentives for self-certification of auditors. Existing compliance-based results could be enhanced by using auditing procedures, and, to ensure confidence in these procedures and audit results, a professional group could certify auditors. Auditors would not have to be third parties. The incentives approach to environmental auditing was also developed by the EPA during this time. However, it became clear that incentives (such as reduced regulator inspections) could not really be promised. The EPA would encounter difficulty in delivering incentives under current environmental laws. Many industries simply wanted the EPA to stay out of environmental auditing.

The final EPA approach resulted in the 1986 publication of EPA's *Environmental Auditing Policy Statement.* This approach can best be summarized using the words "endorsement, analysis, assistance." The policy first states that the EPA will continue to endorse private sector environmental auditing. The EPA examined many industrial examples of environmental auditing programs and decided that the benefits of private programs were substantial. Environmental auditing was to stay in the private sector.

Analysis was also a component of the policy. The EPA was to continue studying the elements and benefits of auditing programs. Finally, the EPA was to assist interested parties in promoting environmental auditing. Perhaps the most significant item in the *Environmental Auditing Policy Statement* is the theme that auditing reports would not, under normal circumstances, be requested by the EPA. However, the EPA did leave open the option of obtaining reports to support enforcement action.

The EPA has occasionally obtained environmental audit results for selected enforcement cases. More frequently, they have used the benefits of environmental auditing in consent decrees. The EPA has required companies, through consent decrees that result from enforcement, to establish and implement environmental auditing programs. Today, the EPA's interest in environmental auditing has given way to interest in pollution prevention programs. No active efforts to promote environmental auditing, through conferences, publications, or other means, remain within the EPA. If anything, the Department of Justice has issued the most recent guidance concerning the use of environmental audits. The EPA is preparing a policy statement on pollution prevention.

Department of Justice Guidance on Factors for Criminal Prosecutions

On July 1, 1991, the DOJ issued *Factors in Decisions on Criminal Prosecutions for Environmental Violations in the Context of Significant Voluntary Compliance or Disclosure Efforts by the Violator*—a legal title for a legal document. The purpose of this document is to attempt to clarify factors for bringing criminal prosecutions for environmental violations without discouraging self-audit programs. Factors for regulators to consider for prosecution action include:

- Degree and timeliness of cooperation by violator;
- Existence and scope of a comprehensive environmental compliance program;
- Within this program, extent of compliance and management audits, including assigned resources committed to auditing and implementation of auditing recommendations;
- Extent of noncompliance;
- Effectiveness of internal disciplinary action; and
- Subsequent compliance efforts.

The DOJ guidance provides several hypothetical examples on the use of this new prosecutorial discretion. Generally, companies with established, well-staffed, and well-funded environmental auditing programs that act on findings and recommendations discovered in an audit may receive mitigation consideration for compliance violations. Discretion is clearly given to the individual prosecutor.

Development of Private Standards and Policies

Government regulation is not the only available mechanism for pollution control. Several professional organizations are involved in the development of environmental auditing standards. The Institute for Environmental Auditing adopted an environmental management program audit standard in March 1992. This standard is not designed to measure compliance. Instead, the standard is more broadly focused by providing procedures for assessing an organization's environmental management program.

The American Society for Testing and Materials (ASTM) is also developing an environmental assessment standard. Although not directly related to air auditing issues, the ASTM standard sets forth procedures for conducting a Phase I environmental assessment for commercial property transactions. Air emissions may be a significant concern for some industrial properties or businesses with extensive vehicle fleets.

IMPACT OF LEGISLATION AND GUIDELINES ON ENVIRONMENTAL AUDITING

The need to audit compliance with environmental regulations, especially for large industries with multiple plant locations, created the initial environmental auditing programs. The newer environmental statutes are forcing environmental managers to consider auditing for performance-based compliance, for beyond-compliance environmental management, and for nonregulated environment, safety, and health issues. Many different types of audits and assessments have been generically considered under the term "environmental audit." These items include:

- Compliance environmental audits;
- Property transaction audits or assessments;
- Waste minimization or pollution prevention audits;
- "Green" audits or total quality management assessments;
- Environmental management program audits;
- Combined environment, health, and safety reviews; and
- Environmental impairment liability insurance assessments.

Of all the auditing types, compliance audits are the best known and are considered to be the traditional environmental audit. More frequently, transaction audits or assessments are conducted for property development projects and commercial property transactions. These studies are designed to identify known or potential issues dealing with hazardous substances on the subject property.

As waste minimization or pollution prevention programs are developed in the government and commercial sectors, audits to measure the progress and performance of these programs will be conducted. Measuring the attainment of a percentage reduction in the emission of toxic air pollutants is one example of this type of audit. Total quality management and "green" concepts are now being applied to environmental issues. Audits to determine compliance with these concepts are being conducted by many large industries.

The concern voiced by financial accounting and insurance professionals over the ability to determine the effectiveness of environmental management programs has led to the development of auditing standards. Similarly, many organizations concerned over the use of separate entities and techniques to audit environment, health, and safety programs have unified E, H, & S auditing under one program. These organizations believe that more consistent audits will be performed by combining these functions.

Finally, risk assessments and environmental impairment liability insurance assessments have combined environmental risk assessment, risk financing, and compliance audit techniques. The result provides risk managers or insurance underwriters with qualitative or quantitative indications of risk. Current and future environmental risks are identified.

All of these forms of auditing have been influenced by legislation, regulations, and guidelines. The following items are among the more important influences:

- Land disposal restrictions, and treatment and disposal standards (RCRA);
- Transportation and storage requirements for hazardous waste (RCRA);
- Effluent limitations and technology wastewater treatment requirements (FWPCA);
- Title III emergency notification and emission reporting requirements (SARA);
- Underground storage tank technical standards and financial responsibility requirements (RCRA and SARA);
- Compliance certification and reporting requirements for air emission permits (CAA);
- Section 8 substantial risk information and health and safety studies (TSCA);
- Voluntary protection programs under the Occupational Health and Safety Administration; and
- Achievement of cleanup levels for past waste disposal practices (RCRA and CERCLA).

Additional legislation is expected to influence environmental auditing. Proposed amendments to the FWPCA and RCRA have focused on hazardous pollution prevention. Under these very similar bills, pollution prevention audits conducted by a third party would be required for selected industrial categories that submit emission inventory forms under SARA Section 313. The legislation also required that the EPA administer training and certification programs for auditors.

These bills have not been passed by Congress, and it is unlikely that the bills will pass with the third-party and certification requirements. If the past is any indicator of the future, one can expect that future legislation will increasingly influence environmental auditing activities. State legislation may take the initiative.

Massachusetts is in the process of establishing a program to license site managers for state cleanup activities. These managers would serve as private parties to administer cleanups under the state superfund program. The managers would oversee all cleanup activities on behalf of the state. Other certification influences are known for property assessments and real estate appraisals. Michigan, Florida, and New Jersey have established environmental property assessment regulations or guidance.

II

Principles and Practices

3

Auditing Techniques

Lisa E. Lowe

INTRODUCTION

The environmental audit field is "continuously evolving and there is no strict or prescribed way to perform an audit" (EPA 1989a, 19). However, there are numerous planning and auditing criteria and techniques that can be used to make an environmental compliance audit program more effective and efficient. The techniques used will be dependent on the audit program's goals and objectives. This chapter describes some of the techniques that can be used when planning or conducting an audit.

AUDIT GOALS AND OBJECTIVES

The first element is to determine the goals and objectives, as these will set the requirements, tone, and boundaries of the environmental audit program. The goals of an audit program can be quite varied and may be incorporated into a company's mission, values, policy, or standard operating procedures. The goals may range from a good corporate citizenship requirement to a more specific definition, such as providing top management with a compliance profile of all operating facilities.

"An environmental audit is a systematic, documented, periodic, and objective review of regulated entities of facility operations and practices related to meeting environmental requirements" (EPA 1986a, 25006). Thus, the primary objectives for conducting environmental audits include:

- Verifying program compliance with applicable permits, regulations, and organizational policies;

- Identifying strengths and/or weaknesses of the existing environmental compliance and/or management programs;
- Identifying current problems, anticipating potential problems, and identifying solutions to respond to such problems;
- Defining responsibilities and accountability of individuals to fulfill legal and business requirements;
- Eliminating or reducing, to the maximum extent practical, environmental pollution resulting from the various manufacturing operations;
- Verifying that facilities are conducting the required corrective actions in response to an organizational audit or an Environmental Protection Agency (EPA), state, or local environmental inspection; and
- Assessing the potential risks and liabilities associated with past and current environmental management programs.

Although outside of the strict sense of an environmental audit as defined by the EPA, other objectives could include communicating new operating practices, identifying ways to reduce environmental program costs, identifying new or additional pollution prevention techniques, or determining additional training needs of environmental personnel.

The goals and objectives that are selected will impact the environmental audit program development and implementation. The seven steps of a typical audit program, as shown in Figure 3-1, include the audit planning task, the preaudit review, the site visit, data analysis, the audit report, the audit action plan, and an evaluation of the audit program. Each of these steps are necessary to accomplish the goals and objectives of an effective audit program.

After the goals and objectives have been established, top management support must be obtained. Support from top management should include corporate officials, attorneys, and environmental personnel, division managers, and plant managers, as well as other pertinent members of management.

The two primary reasons for obtaining management support are as follows. First, within the context of the myriad of environmental laws and regulations, top management has the ultimate responsibility of a company's environmental performance. Thus, top management must understand the size and scope of both the company's and their own potential liabilities under the various environmental laws and regulations. In addition, they are responsible for assuring a commitment to "correct identified problems and prevent their (the problems) recurrence" (EPA 1986a, 25009).

Second, without top management support, an audit program may never fulfill its goals and objectives. For example, the funding and/or other resources required to have an effective program may not be available. In addition, facility personnel may not take the program seriously and may not

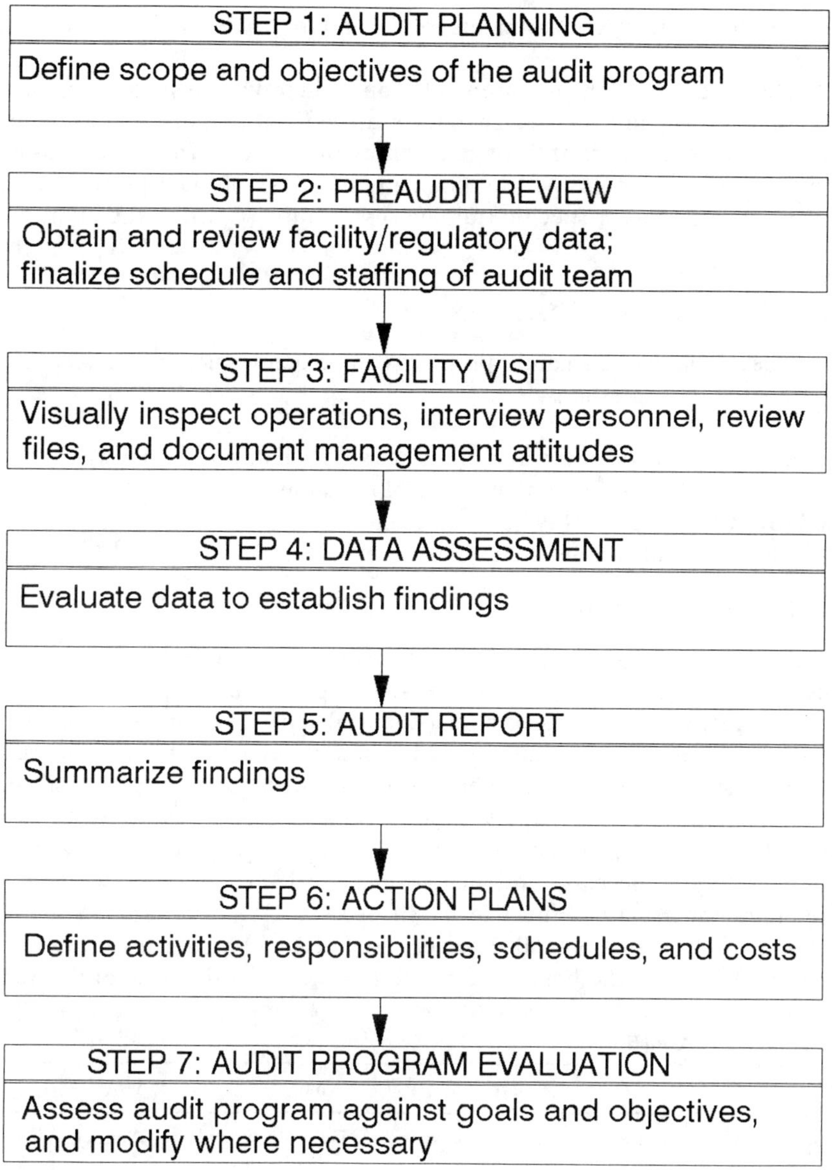

FIGURE 3-1. Audit Steps.

make the necessary commitments to respond to problems identified during the audit.

The need for and the requirement of an audit program should be incorporated into company policies. Then top management support can be communicated to company staff through a variety of methods. These methods can include written memos, videos, company meetings of plant personnel and management, and/or specific training sessions that describe the audit program.

PLANNING THE AUDIT

The development of an audit program can be determined by the responses to several basic questions:

- What is the scope of the audit (defines who, what, where, and when)?
- What are the staffing requirements of the audit team?
- How will confidentiality be maintained?
- What protocols will be used?

Defining the Audit Scope

As can be seen in Table 3-1, once the goals and objectives have been established, the next step is to define the scope of the program. Because an environmental audit requires both a financial and a time commitment on the part of industry, constraints on these commitments will impact the audit scope, the number of audits performed, and who conducts the audit. For example, management may decide to audit all facilities, just those facilities handling chemicals, or just those facilities within a geographical area. Similarly, it may be necessary for management to focus on a particular set of requirements, such as the Clean Air Act (CAA) Amendments or the Resource Conservation and Recovery Act (RCRA) hazardous waste regulations. Whatever the limits placed upon the scope or the scale of the audit

TABLE 3-1 Audit Scope

Who	Where	What	When
Manufacturing facilities	United States	Federal laws and regulations	Annual
Warehouses	International	State laws and regulations	Biannual
Environmental contractors		Local laws and regulations	Past performance
		Corporate policies, standards, guidelines	

program, these issues should be addressed to provide an efficient and cost-effective audit program.

Who

An environmental audit program may be designed to verify compliance of a company's own facilities or may be enlarged to include environmental contractors. If a company is just beginning an audit program, it is best to start off small. Companies may choose to audit all domestic locations first, then their international facilities, and then their contractors.

For most companies, an environmental audit program generally concentrates on their own locations first. However, as a program matures, an increased emphasis should be placed on environmental contractors. The emphasis on contractor auditing is due, in part, to the liabilities that can be associated with their use. For example, under the RCRA, industry has a cradle-to-grave responsibility for the hazardous wastes or hazardous waste constituents that are generated by their manufacturing operations. Because of the liabilities associated with the off-site treatment, storage, and/or disposal of hazardous waste, industry must identify and use those contractors that are in compliance with RCRA and that provide a low management risk.

Companies may audit any off-site laboratories that provide analytical data associated with environmental monitoring requirements, as the company has the ultimate responsibility for the quality of that data. On-site contractors also may be audited to verify that they have an awareness of and are in compliance with the associated environmental requirements.

If a company only has a few locations, it may be possible to audit all locations within a relatively short time frame. However, if a large number of locations are to be audited, then a company may choose to prioritize or rank the audits.

Ranking systems can be based on either a biased or unbiased selection process. Under a biased approach, the locations to be audited first would be selected based on past performance (i.e., those facilities where deficiencies have occurred in the past) or based on risk management decisions.

One example of a risk management ranking system is to base the selection process on the operations conducted at each facility. This ranking assigns a potential risk level of low, medium, and high to the potential release or emission points associated with the operations and to the potential liabilities faced if a release occurs. Because many different operations can result in the same types of emission or release and subsequent liabilities, the rank is established subjectively. A numerical value is assigned to each of the risk levels (e.g., low equals 1, medium equals 3, and high equals 5). Then the various operations are ranked, with those facilities having the highest numerical scores audited first.

A similar biased ranking process could be developed using environmental setting characteristics. In this ranking system, facilities to be audited first would be selected based on their location in relationship to the following:

- Whether the facility is located in an urban or residential setting;
- Whether the facility is located near or adjacent to a public drinking water supply; and
- Whether the facility is located in an environmentally sensitive area, such as a wetlands, a flood plain, or an earthquake-prone area.

Under an unbiased system, facilities would be selected by either a random or a stratified random approach. Random selection is any method based on random chance probability. Under this approach, each facility would have an equal chance of being selected. A stratified random approach would involve the division of the population into groups. For example, selection of the facilities could be on size, location, and/or the number of environmental programs managed by the facility.

What

The areas to be covered in an environmental audit program will depend in a large degree on the goals and objectives of the program, as well as on any financial constraints. For example, a company may choose to focus its auditing efforts on a new law, such as the CAA Amendments, to determine the law's impact on its operating locations and to develop the management programs that will be required to meet proposed regulations.

Financial and time constraints, both on the part of the auditors and the facility to be audited, should also be taken into consideration. An audit program that is designed to verify compliance with all the associated requirements of an environmental regulation will take considerably more effort than a program to define the responsibilities and accountability of environmental personnel. However, the first program would provide management with a much better understanding of its compliance status and any potential liabilities.

Table 3-2 lists the potential environmental laws, and Table 3-3 lists the various areas that may be covered during an audit. In addition, audit programs may be designed to focus solely on or to include the verification of compliance with company policies, standards, procedures, and/or guidelines.

Where

Another important consideration in developing the audit scope is to determine where the facilities are located. Many companies have both domestic

TABLE 3-2 Environmental Laws

Clean Air Act (CAA)
Clean Air Act Amendments (CAAA)
Clean Water Act (CWA)
Resource Conservation and Recovery Act (RCRA)
Hazardous and Solid Waste Amendments of 1984 (HSWA)
Safe Drinking Water Act (SDWA)
Pollution Prevention Act (PPA)
Toxic Substances Control Act (TSCA)
Comprehensive Environmental Response and Liability Act (CERCLA)
Superfund Amendments Reauthorization Act (SARA)
Federal Insecticide, Fungicide and Rodenticide Act (FIFRA)

(United States) and international facilities. The audit programs for these facilities may vary somewhat, as countries may have widely divergent environmental laws and regulations. Many companies choose to audit their international facilities against their own company policies, standards, procedures, and/or guidelines, especially when only a few environmental requirements exist.

Considerations that must be taken into account when defining the scope of international audits include how to obtain current copies of the environmental laws and regulations, how to define the protocols to be used, and how to determine the legal requirements. For example, some countries require environmental audit reports to be filed with governmental agencies. Other

TABLE 3-3 Areas To Be Addressed During an Audit

Air emissions
Wastewater discharges
Stormwater discharges
Hazardous waste
Nonhazardous waste
Underground storage tanks
Aboveground storage tanks
Toxic substances
PCBs
Emergency planning and right-to-know
On-site water supply
Underground injection
Radioactive materials
Asbestos
Chemical storage and handling
Transportation
Chemical/waste

issues to be addressed include potential language barriers, local holidays, travel requirements, and local customs.

Frequency

The scope of the audit program should also include the audit frequency. For small companies, an audit of every facility may be conducted the first year, with a second audit conducted on a less frequent basis. For large companies, audits of all facilities may take several years to complete. Factors that influence the frequency at which audits are conducted include changes in regulatory requirements, changes in the company's activities, major modifications to a facility or its mode of operation, past audit performance, and receipt of notices of violation or administrative orders.

Scheduling the Audit

It is suggested that the yearly audit schedule be published at the beginning of the fourth quarter of the preceding year. All facilities to be audited should be asked to respond within 1 month of any known scheduling conflicts. The early posting of the audit schedule will allow facilities to identify any problems with the audit dates and allow the audit schedule to be amended, where possible.

The Audit Team

The next step in the audit-planning task is to determine the staffing requirements of the audit team (e.g., size, qualification). The auditing personnel should be sufficiently versed in the various state, federal, and international regulatory requirements, auditing protocols and techniques, risk management techniques, and the equipment and procedures that can be used to meet regulatory requirements.

The audit team can consist of either industry personnel and/or outside consultants. The advantages of using internal auditors are that it reduces the potential for outside disclosure of data and the personnel assigned to the audit are more likely to be familiar with the operations conducted at a facility. In addition, it may be easier to revise audit schedules. The disadvantages of this approach are that in-house staff may not be trained to perform environmental audits and also may not be able to provide unbiased opinions concerning the degree of environmental compliance. In addition, using internal personnel may increase human resource requirements (each audit can take between 120 to 300 hours to complete).

The advantage of using external auditors is that they provide an independent, unbiased view of the environmental performance at a facility. Other advantages are that additional personnel may not need to be hired and external personnel tend to have more of the necessary prerequisite skills. Disadvantages include possible disclosure of data, potential scheduling and communication problems, and a lack of familiarity with company operations.

Most companies use a variety of approaches for staffing environmental audits. Audits are often performed in-house, using either corporate or technical personnel from other facilities. For formal audit programs, many companies establish an independent internal auditing group, hire outside consultants, or adopt a composite approach, using both internal and external personnel.

Based on the need to protect the confidentiality of the data obtained during an audit, many companies also choose to include a lawyer as part of the audit team. "Confidentiality cannot be accomplished without the active involvement of legal counsel, because the means for protecting information from disclosure are the attorney-client privilege and the attorney work product privilege" (Ryan 1991, 5).

Confidentiality and Disclosure

It is necessary to be aware that sensitive information concerning compliance with environmental laws and regulations will be generated, as well as information concerning trade secrets. Although the EPA has indicated in their policy statement that they will limit their requests for audit reports, they do have the statutory authority to request all relevant information on a facility's environmental compliance status. To this extent, it is necessary to determine in the planning phase how confidentiality will be handled.

There are two primary mechanisms that are used to protect audit confidentiality—the attorney-client privilege and the work product doctrine. However, neither the attorney-client privilege nor the work product privilege will prevent regulators or prosecutors from discovering audits that expose compliance problems. Some companies no longer use either of these mechanisms.

If a company selects to use either the attorney-client privilege or the work product privilege to protect confidentiality, this must be discussed with legal counsel prior to beginning the audit program. This is because there are many requirements that must be met to prove that either of these privileges exist. For example, the attorney-client privilege only protects confidential communications between an attorney and client when a set of specific requirements are met. This protection extends only to communication and does not prevent disclosure of underlying facts.

The second mechanism, the attorney work product privilege, only protects materials prepared by or for counsel *in anticipation of litigation for trial.* "Again, work product material may be discovered if the party seeking disclosure is able to show that there is a substantial need for the material and he will experience undue hardship in obtaining that material through other means" (*Hickman* v. *Taylor,* 329 U.S. 495 (1947); *c.f.*, Rule 26 (b) (3) of the Federal Rules of Civil Procedure as reported in Ryan 1991, 7).

An attorney also determines the procedures that will be used for disclosing data obtained during the audit to the appropriate governmental agencies. This is because voluntary disclosure of auditing findings does not automatically protect a company from penalties. A recent Department of Justice document provides insight on how the department will respond to compliance data obtained during an environmental audit. This document, which was issued on July 1, 1991, is titled *Factors in Decisions on Criminal Prosecutions for Environmental Violations in the Context of Significant Voluntary Compliance or Disclosure Efforts by the Violator.*

The attorney must determine if there are any regulatory requirements for disclosing data. For example, under the CAA Amendments, "the amendments convert the knowing violation of virtually every requirement into a felony crime" (Quarles and Lewis 1990, 54). Fines for individuals can range up to $250,000, with fines for companies ranging up to $500,000. "However, when the CAAA was passed, Congress included an explanatory statement in the legislative record that, although not part of the actual statute, they did not intend to discourage audits. In fact Congress encouraged companies to audit themselves" (Wunderli 1991, 13) and said that "criminal penalties should not be applied where one, acting in good faith, promptly reports the results of an audit and promptly acts to correct any deviation" (Congressional Record 1990, H13197).

Audit Protocol

There are as many different audit protocols as there are audit scopes. The protocol can consist of a checklist, a questionnaire, a formal document describing the procedures to be used during the audit, or a combination of the above. Whatever format is used for the protocol, it must be sufficient to meet the goals and objectives of the environmental audit program. For example, a simple summary checklist may be used to focus the auditor's activities, to identify the records and documents to review, or to obtain information that is not systematically written down. Summary checklists are often used for international audit programs or for auditing against company standards and guidelines.

If an environmental audit program is to be used to verify compliance with environmental regulations, a very detailed questionnaire may be used. Typically, questionnaires consist of a series of yes/no questions, with the regulatory requirements used as a basis for preparing the questions. The questions or items on the questionnaire are designed to determine if the facility's procedures and practices are in compliance with the selected environmental regulation.

It should be noted that, with constant changes to the regulations, a questionnaire can become outdated in a matter of months. Therefore, the questionnaire should list the date of preparation so that the auditors can more easily identify additional regulatory requirements not addressed in the questionnaire. The questionnaire may also be updated prior to conducting each audit, on an annual basis, or a company may choose to use one of the computerized questionnaires that have been developed by several consulting firms.

A formal auditing protocol consists of a written document that describes the goals and objectives of a company's environmental audit program, as well as each of the program's elements. The use of a formal, documented audit protocol ensures consistency in the audit program. It can also be used as a training device. The audit protocol manual may include specific auditing techniques that the auditor is to employ during the course of the audit, copies of the checklists or questionnaires, a section on auditing dos and don'ts, and the report format.

PREAUDIT ACTIVITIES

The purpose of the preaudit task is to confirm the audit schedule, define the resources needed to conduct the site visit and to prepare the audit report, familiarize the audit team with the facility operations and regulatory requirements, and determine the data gaps to be addressed in the site visit and the relative degree of compliance.

Approximately 2 to 3 months before the audit, the audit team leader confirms the audit dates with the facility. A follow-up letter is sent to the appropriate facility personnel. This letter should include the confirmed audit dates, a list of personnel who will be conducting the audit, the objectives of the audit, the audit scope, and a tentative agenda. At this time, the facility should also be requested to provide data. The information to be obtained prior to the audit may depend on the type of or size of the facility, but at a minimum should include the following:

- A site map, including physical layout;
- A general description of the facility, including the operations conducted, chemicals used, and products made;

- Any current or future planned process modifications;
- A list of personnel responsible for environmental/safety programs, engineering, production, materials management, and human resources;
- A copy of any previous audit reports, self-audit reports, and/or inspection reports from local, state, or federal agencies, including copies of facility action plans; and
- A copy of any notices of violation or administrative orders.

If the facility is quite large or if there are significant time or personnel constraints, the team leader may request the facility to provide copies of additional records, including:

- Emission inventories;
- Discharge inventories;
- Pollution control equipment specifications;
- Permits;
- Monitoring reports;
- Facility plans (e.g., emergency response and waste minimization plans);
- Training records; and
- EPA- or state-required reports (e.g., annual or biennial hazardous waste reports, Tier I and II reports, Toxic Release Inventory Reports).

In addition, the request may include a simple checklist for the facility to complete to determine those areas where compliance is an issue. An example page of a checklist is provided as Exhibit 3-1.

Regulatory agencies with jurisdiction over the facility are identified during the preaudit activities. For example, if a state has been granted authority to fully administer a regulatory program, the state regulations are obtained. If a state only has partial authority or administers its own program in addition to the federal program, then both state and federal regulations are obtained and the differences between the programs are defined. In addition, any local agencies, such as a county health department, may also have additional requirements that a facility must meet. These requirements are identified, and copies of the pertinent local laws and regulations also are obtained.

The preaudit review determines the resources needed to conduct the site visit and identifies the technical skills that the auditors should have to perform a complete investigation. The team leader should define each audit team member's responsibilities during the audit. The team leader will also provide each team member with the materials obtained from the facility, copies of a list of the regulatory requirements that will need to be reviewed, and an audit agenda.

Air Emissions:

A. Applicability
1. Are there any stacks, vents, hoods, or pipes from the facility to the atmosphere? ____ Yes ____No
2. Are any chemicals, vapors, gases, or particulate matter released from this facility through stacks, vents, hoods, or pipes? ____ Yes ____No
3. Has the facility received complaints from neighbors regarding odors? ____ Yes ____No
4. Has the facility received any notice of violation or corrective actions regarding the need to obtain or modify air permits? ____ Yes ____No

B. Emissions Inventory
1. Has a facility-wide air emissions inventory been conducted? ____ Yes ____No
2. Have emissions and operating data for each source within this facility been inventoried? ____ Yes ____No
   a. Stack data (height, diameter, temperature, velocity)? ____ Yes ____No
   b. Pollution control equipment data (type, efficiency)? ____ Yes ____No
   c. Emission rates (annual, hourly maximum, hourly average)? ____ Yes ____No
   d. Operating schedule and capacity? ____ Yes ____No
3. Have emissions data been updated and submitted to the state, if required? ____ Yes ____No

C. Permits
1. Does the facility have any state or local permits for any of the air emission points? ____ Yes ____No
2. Is there a management system for determining permit needs for new or modified sources? ____ Yes ____No
3. Is documentation of air permit requirements available in the facility files? ____ Yes ____No

EXHIBIT 3-1. Example of an Air Emissions Checklist.

For small facilities with minimal operations, only one auditor may be required. However, for larger facilities, two auditors are required for approximately 2 to 3 days. (This does not include legal counsel that may be present as part of the program's confidentiality requirements.) The use of two auditors is especially important if a number of environmental programs are to be reviewed. For example, one auditor may concentrate on hazardous waste requirements, while the other auditor investigates the facility's air emission programs. This technique is very effective in that it allows the work to be divided based on the auditor's area of expertise. The work also may be split to allow one auditor to review facility records (especially those not available prior to the site visit), while the second auditor interviews facility personnel.

For more complex sites or those sites with numerous environmental programs, additional personnel may be required or a longer time may be needed to conduct the audit.

SITE VISIT

A site visit reinforces management's commitment to an audit program and lends support to the personnel charged with the responsibility of a facility's environmental programs. The site visit is used primarily to address data gaps, inspect operations, review in-house files, and determine management attitudes toward the environmental programs.

The agenda of a site visit consists of a number of activities, including:

- An initial meeting of the audit team with the facility staff;
- A brief tour of the site to familiarize the audit team with overall plant operations;
- A detailed site inspection and file review;
- A presentation of daily findings to facility staff; and
- A final close-out meeting.

The initial meeting is often the first opportunity for the audit team members and the facility staff to get acquainted. The audit team should try to integrate themselves among the facility staff rather than sitting together to help establish rapport and to eliminate the projection of an "us against them" position. During this meeting, the audit team leader describes why the audit is being conducted, details its goals and objectives, and goes over the audit agenda. The facility is requested to provide a description of the facility operations and the safety procedures that must be followed by the audit team.

The initial facility tour is brief and is used to orientate the auditors and to help them determine which areas should be examined in depth. The tour includes an inspection of all process operations, all chemical/waste storage and handling locations, pollution control devices, discharges to the atmosphere, land, or water, and sometimes a perimeter survey of the facility. It is the team leader's responsibility to keep the audit team moving.

The information obtained during the initial meeting and the tour of the operations is used to prepare for the remainder of the site visit. This data may require the team to reevaluate which records will be reviewed and what areas to focus on, based on the initial visual observations. A meeting may be held to determine which areas each auditor will review in depth.

The three methods by which audit findings are developed during the site visit are through visual observations, a review of files, and interviews with facility personnel. To establish a record of the audit findings, auditors can use a bound, dedicated notebook to record their observations and to file review notes and interview notes. Pertinent information, such as the name of the person interviewed, the name of the report that was reviewed, and the discrepancies that were noted, should be recorded. Only facts should be recorded, not conclusions. Depending on the audit program requirements, this notebook may be destroyed at the end of the audit, so only the final report remains as a permanent document.

Because of the complexities of the regulations, it usually is not possible to review all records or documents. Therefore, the audit protocol usually defines the system by which records are reviewed and observations verified. The methods used to select which records will be reviewed typically include both random and nonrandom sampling techniques. "The purpose of random sampling is to collect information which can be used to extrapolate or make inferences about the general population being sampled. Nonrandom sampling may be chosen when information regarding specific points is more important than general inferences about the population at large" (EPA 1986b, 10–25). "Regardless of the sampling method(s) used, it takes judgement both to evaluate the soundness of internal controls and to interpret the significance of the results obtained from verification testing" (Greeno, Hedstrom, and DiBerto 1987, 122).

At the end of each day, a meeting is held with facility staff to discuss the daily audit findings and to clarify any questions. The next day's agenda is also discussed. The auditors describe the additional areas they are planning to investigate, additional records to be reviewed, and the personnel they would like to interview. This allows the facility to determine when personnel will be available and allows the auditor to appropriately schedule their activities.

On the last day of the audit, a close-out meeting is held. Sufficient time (approximately 1 to 3 hours) should be set aside for this meeting. The purpose of this meeting is to describe and, where possible, to obtain agreement on the audit findings. Additional objectives of the meeting, depending on the audit scope, may be to assist the facility in developing recommendations, their action plan, or cost estimates.

DATA EVALUATION

Analysis of the data collected prior to and during the site inspection is necessary to evaluate compliance. Information concerning facility management programs and procedures, permits, and records will have been collected. This information is reviewed so that the findings on the facility's past procedures, present activities, and any future planned activities can be prepared. The auditor should analyze each finding to determine if it answers the questions who, what, when, where, why, and how. In other words, is the finding well documented or is further inquiry necessary?

When the audit program focuses on environmental compliance, data evaluation consists of a comparison of facility operations, programs, and procedures to regulatory requirements, including all administrative, technical, and management requirements. When an audit program is used to identify the strengths and weaknesses of management programs, data evaluation will focus on an analysis of the management systems and procedures. For those audit programs that include a risk management objective, the evaluation step will also include an assessment of the potential risk and liabilities associated with a facility.

Administrative requirements include the permits, reports, plans, records, and any other administrative actions required by an environmental regulation or a company standard or guideline. Although a review of administrative requirements may be quite time consuming, it generally requires less skills on the part of an auditor.

Technical requirements include, but are not limited to, the effectiveness and implementability of facility plans, the efficiency of pollution control devices to meet permit requirements, the compatibility of tanks and tank contents, the adequacy of secondary containment systems, and the adequacy of sampling locations, methods, and/or analytical procedures.

Management requirements include any standard operating procedures, guidelines, or plans that describe how a facility conducts its environmental programs.

In evaluating risk to public health or the environment, several fundamental factors are considered, including:

- The relative toxicity of the materials or waste that potentially could be released; and
- The presence of a complete exposure pathway (i.e., is there a source or mechanism of release, a transport medium, a point of contact with the transport medium, and an exposure route).

Data that is evaluated includes operating characteristics, compliance history, monitoring data, visual evidence regarding emissions or releases, process characteristics and residue controls, and the facility setting.

AUDIT REPORT

It is important to establish the type of report and the format that will be used so that the audit findings are reported in a consistent manner. A summary of the elements of a good audit report can be found in the EPA's *Fundamentals of Environmental Compliance Inspections* (1989b, 17-8). As described in this document, "fairness, accuracy, completeness, conciseness, clarity, and organization are all essential characteristics of well-written and effective inspection (audit) reports." Each of these elements are important to the overall content and should be used while writing and reviewing the report.

Three report types can be used to summarize the findings of the audit—an oral report, a written report, or a computer-generated report. For most companies, an oral report is used only at the site visit close-out meeting. This provides the facility as well as the audit team the opportunity to resolve any outstanding issues or questions and to agree to the audit findings. However, an oral report does not take the place of a written or computer-generated report.

There are many different styles of written audit reports. However, there is general agreement that an audit report should include the following elements at a minimum:

- An executive summary, which summarizes the major findings for top management;
- An audit description, which includes the date of the audit, the auditors' names, the operations audited, and the procedures followed; and
- The audit findings, which include a factual finding with the corresponding regulatory citation.

Some reports include a narrative description of the observations made, the records reviewed, and the individuals interviewed as part of the audit finding or as a backup to the finding. The audit report may also include copies of the checklist or questionnaire used to conduct the audit. In addition, audit reports sometime include the auditor's recommendations for remedying the finding.

Each of the findings presented in an audit report must be a statement of fact. They should not include opinions or draw legal conclusions. Also, they should be based on hard evidence obtained during the preaudit and site visit activities.

Although sometimes all findings are included in an audit report, an important technique is to determine if the findings are related to a management system that needs to be corrected. For example, a finding that 1 drum out of 300 was not labeled would not necessarily indicate a problem with the facility's drum management system. However, if 40 drums out of 300 were not labeled and 60 of the labeled drums did not contain the data required by facility's policy, the finding should focus on the management system rather than the individual components.

ACTION PLANS

No matter how well planned or how successfully conducted, an audit program is not an effective management tool unless facilities are required to take corrective action with regards to the audit findings. The response that is received from the facility should include the strategy for responding to the audit finding. The strategy must include a description of the activities to be undertaken, who is responsible for the activity, the activity milestones, the projected activity completion date, and an estimated cost to complete the activity.

The action plans should define the necessary changes in the facility operations to achieve compliance with applicable environmental laws and regulations or with company standards. In some cases, compliance may be easily rectified by changing a facility's operating or recordkeeping procedures. In other cases where lack of compliance may be due to inadequate control technologies, engineering studies may have to be performed, systems designed, and control equipment installed and maintained to ensure that regulatory requirements are met.

Although there should be concurrence with each of the audit findings by the end of the close-out meeting, it is possible for a difference of opinion to occur. In that case, the facility should describe why the audit finding is incorrect and/or the reasons why no action is required. The ultimate resolution of differences of opinion will depend on who is responsible for reviewing the facility's response.

Often, the plant manager and/or his or her superiors are requested to sign the formal response to assure management's commitment to the action plan. The action plan is then submitted for review.

It is important for the review process to be defined so that a method for resolving differences of opinions regarding audit findings is established. This will help to assure that the facility's activities will result in the desired action. The review of the draft audit report may be conducted by corporate staff, an internal audit department, a consultant, and/or the audit team members. Based on the review, the facility may be required to revise the action plan before it is finalized.

A formal system to document the facility's implementation of the action plan should be established. This system may include preparing a monthly, quarterly, biannual, or annual report to the audit department, corporate staff, or legal department. This report should be reviewed by the appropriate personnel. Progress made in responding to each of the audit findings is noted. If schedules have slipped, the facility is requested to provide an explanation of the schedule delays and what corrective measures will be implemented to assure that the action is completed on time.

During subsequent audits, auditors should give special attention to the previous audit reports and action plans. This provides an additional technique to determine how the facility has responded to their action plan.

EVALUATION OF AN AUDIT PROGRAM

An evaluation to determine if the audit program is meeting its goals and objectives is the final step of the audit program. This step assures companies that any identified problems can and will be corrected.

For companies just starting an audit program, one approach is to conduct one or more trial audits to address any unforeseen problems. This will allow companies to modify their program before launching it on a full-scale basis.

An initial audit exercise may identify changes to be made in the audit protocol, changes in the audit report format, changes in the size or composition of the audit team, and/or the need for additional training. Once a program has been established, it is suggested that it be reviewed on an annual basis. This review can significantly improve the quality of an audit program and may require a company to revise its goals and objectives.

An evaluation of an audit program can be accomplished by eliciting responses from auditors, from personnel responsible for monitoring action plans, and from management, as well as from the facilities that were audited. The responses that are received are reviewed, and the program is modified, where appropriate.

Another evaluation technique is to review the audit reports and action plans produced during the preceding year. The review is used to determine if the audit findings meet the goals and objectives of the program (e.g., is compliance being verified and are the appropriate actions being taken by the operating facilities). This review typically is conducted by internal personnel not involved in the audit program or by external personnel.

CONCLUSIONS

A well-defined audit program can help you identify and manage your risks. Although not all techniques to plan and conduct an environmental audit program have been described within this chapter, it is important to remember that the goals and objectives will drive the program. Therefore, sufficient planning must be implemented before beginning a program to make it effective. Planning must include defining the goals and objectives, obtaining top management support, and determining the scope of the audit, who will conduct the audit, and how confidentiality and disclosure will be handled. A protocol must be developed to guide the auditors during the preaudit and site visit activities. A reporting format must be established, and facilities must be responsible for preparing and implementing corrective action plans. Finally, the environmental audit program must be reviewed to determine if it meets the goals and objectives that were originally established.

References

Congressional Record, October 26, 1990. Joint Explanatory Statement of the Committee of Conference.

Environmental Protection Agency. 1986a. "Environmental Auditing Policy Statement." Federal Register. 51(31).

Environmental Protection Agency. 1986b. *Draft Data Quality Objectives Development Guidance for Uncontrolled Hazardous Waste Site Remedial Response Ac-*

tivities. OSWER Directive 9355.0-7A. Washington, D.C.: U.S. Environmental Protection Agency.

Environmental Protection Agency. 1989a. *Environmental Audit Program Design Guidelines for Federal Agencies*. PB90-138744. Washington, D.C.: U.S. Environmental Protection Agency.

Environmental Protection Agency. 1989b. *Fundamentals of Environmental Compliance Inspections*. Rockville, Maryland: Government Institutes, Inc.

Greeno, Hedstrom, and DiBerto. 1987. *Environmental Auditing Fundamentals and Techniques*, 2nd Ed. Massachusetts: Center for Environmental Assurance, Arthur D. Little, Inc.

Quarles, John, and William H. Lewis, Jr. 1990. *The New Clean Air Act: A Guide to the Clean Air Program as Amended in 1990.* Washington, D.C.: Morgan, Lewis, & Bockius.

Ryan, James E., Jr. 1991. Role of outside counsel in environmental audit. Paper read at the 84th Air & Waste Management Association Annual Meeting & Exhibition, 16–21 June 1991, at Vancouver, British Columbia.

Wunderli, Earl M. 1991. Multi-national corporate strategy for environmental assessments. Paper read at the 84th Air & Waste Management Association Annual Meeting & Exhibition, 16–21 June 1991, at Vancouver, British Columbia.

4

Quality Assurance for Environmental Auditing Programs

Michael D. Henke and Jill Barson Gilbert

INTRODUCTION

Although many companies adopted and implemented environmental auditing programs throughout the last 15 years, renewed interest has developed lately as a result of the U.S. Justice Department's issuance of guidelines for enforcing environmental violations. Under these guidelines, a company's detection and prompt correction of environmental violations can reduce or eliminate criminal liability. Environmental auditing remains one of the best ways for detecting noncompliances. Combined with strong follow-up programs and the commitment of management, they can correct environmental deficiencies.

Many of the environmental auditing programs originally were developed under different political circumstances. Over time, the Environmental Protection Agency (EPA), state regulatory agencies, local governments, and foreign governments have continued to develop more stringent regulations. The general public, customers, employees, and shareholders have developed a very different perspective about the environment, and their expectations about corporate behavior have grown. As a result, industrial companies are now held to a higher level of responsibility. It is a constant challenge for companies to ensure that their operations are meeting or exceeding these changing requirements and expectations.

Environmental audit programs typically seek to evaluate compliance status against legislation, regulation, company policies, and industry standards. Some programs also include evaluating operations against good operating practices to manage risks. The most proactive of programs also evaluate the systems in place to manage environmental issues and maintain compliance. The audit programs also communicate findings to a level of the organization where action can be implemented to achieve compliance or improve the operation.

Quality has been the ubiquitous buzzword for American industry throughout the late 1980s and early 1990s, and shows no sign of letting up. Long-time quality advocates such as W. Edwards Deming and Philip Crosby have spawned legions of disciples, each presenting their own philosophies or formula for quality, with applications in manufacturing and service sectors. The need for quality and quality assurance in environmental audits is every bit as critical as in any other sector.

The objectives and exact design of the audit program vary according to cultural and political factors, management philosophies, and public expectations. As these factors change, how can senior management assure itself that older environmental audit programs have not settled into a comfortable but outdated niche? How do companies assure that the audit programs are dynamic enough to adjust to changing needs? Finally, how do companies assure that the programs are implemented effectively, consistently, and thoroughly? Quality assurance mechanisms can help meet those needs. The remainder of this chapter illustrates quality assurance of environmental audit programs, quality assurance mechanisms, conducting quality assurance reviews, and components of environmental audit programs.

QUALITY ASSURANCE OF ENVIRONMENTAL AUDIT PROGRAMS

Auditing requires a cumulative summary of dozens of tasks, decisions, assumptions, calculations, and conclusions. In a limited sense, quality assurance is a means for companies to assure the accuracy and thoroughness of environmental audits, maintain the quality of the audit system itself, and provide assurance of consistency and reliability. In a broader sense, it also provides an opportunity for a reality check to assure that the audit program objectives are appropriate.

A number of prominent organizations have endorsed using quality assurance in environmental auditing programs:

- The EPA, in its 1986 Environmental Auditing Policy statement, listed quality assurance as one of seven general elements likely to be included in an effective environmental auditing system.
- The International Chamber of Commerce endorsed quality assurance in a 1989 position paper. Quality assurance is listed as one of seven essential elements of effective environmental auditing that enable the audits to yield maximum benefits.

- The Environmental Auditing Roundtable, in its 1990 Guidelines for Written Environmental Audit Program Procedures, endorses the use of quality assurance mechanisms as 1 of 13 core procedures that should be included in formal environmental audit programs.

QUALITY ASSURANCE MECHANISMS

Although the design of environmental audit programs can vary substantially in response to the needs and culture of a company, it is important to ensure that, once the program has been designed, it is effectively implemented and managed. This quality assurance can be provided through a variety of mechanisms, such as external reviews and certification, internal reviews, and management of the audit process. Management techniques and controls can be applied continuously, while reviews usually are conducted on a periodic basis.

Design and Management of the Audit Process

Design and management of the audit process entails the application of several management functions, such as planning, organizing, staffing, and controlling. Control of the audit process typically involves establishing a set of standards, measuring and evaluating performance against these standards, and correcting negative deviations. The environmental audit program can be designed with quality assurance and control mechanisms. An advantage of these mechanisms is that they can be used on an ongoing basis, but they are limited in the sense that they are primarily inward-looking mechanisms. These approaches are very good at judging the conformance of the process to standards, but they are not very effective at judging whether the standards are valid. Nor are they very useful for judging the compatibility of the program with its objectives.

Standards for conducting environmental audits usually are outlined in protocols and procedures. Establishing and implementing clear, detailed protocols and procedures can help ensure the thoroughness, accuracy and consistency of an environmental audit program. The protocols should match the objectives and scope of the program, and they should cover all phases of the audit process, from selection of sites for audits through management of audit documentation. Checklists developed to support the protocols and procedures should address all applicable environmental topics to an appropriate level of detail.

Direct supervision of the field audits is one of the best ways of assuring that protocols are followed on a consistent basis. It also can provide valuable feedback for the primary customer of the program and can be used to identify improvements for the program.

In its simplest form, supervision involves directing the audit program and the activities of audit team members to achieve the best possible performance. In the context of an environmental audit, performance usually means accurately identifying noncompliances, thoroughly evaluating them, and clearly and concisely documenting them in an objective fashion. Some audit programs also require audit teams to recommend corrective actions.

Most companies conduct audits with teams of individuals selected from within or outside the organization. Each team has a leader. The primary role of the team leader is to organize the audit team and supervise its activities so that it effectively accomplishes the objectives of the program. Leaders for audit teams usually are selected on the basis of skills, knowledge, experience, and training. They should also be selected for organizational and supervisory skills because these traits are required to enhance the functioning of the team.

Team leaders should be able to communicate effectively and analyze technical issues. They should know and understand regulatory requirements as well as the operation of the facility being reviewed. Experience is an important factor that enables a team leader to earn the respect and credibility of the facility being audited and the audit team. Training also is required to ensure a good understanding of the audit process, the objectives of the program, and the skills necessary to collect and analyze information.

In larger companies, it is not practical to have one individual lead all audit teams because of the length of time involved and because the repetitive nature of the work can lead to burnout. However, when more than one person is involved as a team leader and when teams are staffed with a number of people from throughout the organization, there is an increasing potential for inconsistent interpretations or application of the audit protocols.

To reduce inconsistencies, most companies staff audit teams with experienced auditors or provide training for the auditors and team leaders. The objective of the training is to increase awareness, improve skills, and develop consistency. The training focuses primarily on program objectives and procedures, auditing skills, and audit standards.

Another procedure used to reduce inconsistencies is peer reviewing of reports. Companies often will ask team leaders who are not involved in a particular audit to review and comment on an audit report. In this way, differences of opinion or approach can be identified and resolved before issuing a written audit report.

An additional procedure used to evaluate the effectiveness of environmental auditing programs is the survey. Under this procedure, managers whose facilities recently have been the subject of audits receive and complete a short survey regarding the conduct of the audit. This method provides important feedback and can be particularly useful if one of the main objectives of the audit program is to assist facility management. Surveys will typically ask for evaluations of the professionalism and knowledge of the audit team, the timeliness and clarity of the audit report, and the overall usefulness of the audit. A survey also can be used to solicit ideas for future improvement to the audit process.

Internal Quality Assurance Reviews

Periodic internal reviews often are conducted by companies to evaluate the performance of various functions. These may be operational reviews, quality improvement reviews or financial reviews. Often they are performed by a formal audit group. Otherwise they may be performed by management or by an ad hoc committee.

Internal reviews also can be useful for determining the conformance of audit programs with protocols and procedures, and they can help evaluate the compatibility of program objectives with program design. However, internal reviews do not provide independent, objective opinions.

One of the most popular ways of internally reviewing environmental audit programs is through an oversight program. Oversight usually is provided in large corporations, where the diversity of operations precludes the implementation of one standard environmental audit program. In these situations, each business group typically develops and operates an audit program tailored to its own operations. Nevertheless, there is still a need to oversee the operation of these programs, to ensure satisfying the needs of corporate management.

The corporate audit oversight group provides general guidelines for designing the individual audit programs, and conducts reviews to confirm effective implementation of the program. But the oversight group makes no attempt to audit the compliance status of any individual facilities.

The audit program guidelines are designed to ensure meeting the corporate needs while providing enough flexibility for the business group to meet its unique needs. Specific guidelines typically are established to set minimum standards for the scope, organization, staffing, scheduling, and documentation of audits. Guidance may also be provided through standardized checklists or workbooks.

Periodic reviews of the audit programs are performed by the oversight group to ensure conformance with the guidelines and effective implementation. These reviews include interviews with the program manager, the audit staff, team members, and management, to gain insight into the operation and usefulness of the program. Documents, such as checklists, protocols, and reports, are reviewed for content and clarity. A thorough oversight review also requires the observation of a sample of audits, to confirm that they are conducted according to the protocols.

An alternative method of internally providing quality assurance is through conducting multiple layers of audits within a company. This approach also arises in large corporations, where it is important to meet the needs of facility, divisional, and corporate management. Under this approach, the corporate office operates an independent environmental audit program, to achieve as much independence as possible and to maintain a direct line of communication between auditors and management. The corporate program visits facilities and evaluates their compliance status. Even though the primary focus of the corporate audit program is not to evaluate the effectiveness of the divisional audit program or the facility self-audit program, comparisons between the findings of the various programs are almost inevitable. These comparisons can serve as a form of quality assurance.

External Reviews and Certification

Quality assurance also can be provided for environmental auditing programs through external reviews. These reviews normally are conducted by firms who provide environmental auditing services commercially. Special-purpose quality assurance reviews also are conducted by law firms.

Like management controls, external reviews typically provide an analysis of how effectively and consistently the audit program is being implemented. Like internal oversight reviews, they can be used to evaluate the appropriateness of the program objectives and to judge the ability of the audit program to meet its objectives. An important advantage, however, is that external quality assurance reviews can provide an independent opinion of the audit program and can compare the program against similar programs used by other companies.

External quality assurance reviews are conducted in a manner similar to the internal oversight reviews. They consist of interviews, document reviews, and observation of field audits. Findings and recommendations of external quality assurance reviews are documented in a written report. Some companies conducting these reviews also will provide a certification regarding the quality of the audit program. The certifications typically indicate that the program uses state-of-the-art auditing techniques and that it is among the leading environmental audit programs in a particular industry.

CONDUCTING QUALITY ASSURANCE REVIEWS

A very important distinction exists between the actual environmental compliance audit program and the quality assurance audit that evaluates it. In the environmental audit, it is important to focus on the actual status of the facilities, with regard to specific regulatory or company requirements. In the quality assurance audit, it is important to focus on the systems in place to accurately and thoroughly conduct environmental audits on an ongoing basis.

The primary objectives of a quality assurance audit are to confirm that:

- The objectives of the environmental audit program are appropriate.
- The audit program is adequately designed to accomplish the objectives.
- Audits are effectively executed to address the needs of plant and company management.
- Adequate resources and a well-trained staff are assigned to the program.
- Findings of the audits are communicated to the appropriate levels of the organization in a timely manner and are adequately protected to minimize liabilities.
- Corrective actions are implemented on a timely basis.

The scope of work for a quality assurance audit typically includes consulting with key personnel throughout the organization, reviewing documentation from previous audits, and observing current audits as they are conducted.

Interviews

While conducting a quality assurance audit of an environmental audit program, it is essential to interview a broad cross section of people within the company. Senior-level managers can provide background about the company's strategic goals, its commitment to environmental stewardship, and the objectives of the audit program. The environmental manager or the audit manager can provide information about the scope, methodology, staffing, and implementation of the program. Important data about the number of audits conducted, the timeliness of reports, and any significant problems with the program also can be discovered through interviews.

It is also useful to discuss the program with team leaders, team members, and plant managers who have been involved with past audits. Feedback from plant managers is a very important way to determine how effectively past audits have been conducted and how well they have assisted the operating facilities in their day-to-day compliance responsibilities.

Finally, these interviews are also an important way to solicit suggestions for improvements from those people who are closely affiliated with the program.

Document Reviews

A review of documents prepared for the audit program also can provide a useful look at key elements of the program. A written description of the program should provide the charter and purpose of the program. It also should explain the objectives, scope, organization, staffing, and methodology. Written protocols and checklists will reveal the depth of the audits and the diligence with which the audits are conducted. Recently completed audit reports will show how findings and recommendations are communicated throughout the organization. Action plans explain how the company responds and follows up on the audits.

Following is a list of documents that typically define an environmental audit program and that should be reviewed as part of a quality assurance audit:

- Annual report;
- Corporate organization chart;
- Environmental policies;
- Audit program description and charter;
- Protocols and procedures;
- Checklists and questionnaires;
- Training manual;
- Audit reports;
- Action plans/follow-up reports; and
- Periodic status reports.

Observation of Audits

The third element in the scope of a quality assurance audit is participating in and observing actual audits. This element provides a close look at the orientation and training provided for team members, the level of preaudit preparation, the teamwork involved in conducting the audit, the depth of investigation into the environmental topics, the communication of audit findings, and the development of action plans.

COMPONENTS OF ENVIRONMENTAL AUDIT PROGRAMS

The quality assurance audit should evaluate all important components of an audit program. The key components include:

- Objectives;
- Scope;
- Methodology;
- Staffing;
- Organization;

- Documentation of findings;
- Follow-up mechanisms; and
- Measures of performance.

This section considers each of these components in more detail.

Objectives

It is critically important that the objectives are consistent with the company's culture and strategic business objectives. The objectives of an audit program must be clearly understood and communicated throughout the organization if the program is to be effective. It is usually best if top management support is demonstrated by a written directive or policy that articulates the company's commitment to compliance, risk reduction, and other objectives. If the objectives of the program are not clearly defined or are ineffectively communicated, audits often are conducted inconsistently and incompletely, it is difficult to obtain the necessary internal resources on loan from other areas of the company, and there is little real improvement.

Typical primary objectives include assurance of compliance with regulations, policies, and standards, evaluation of hazards, reduction of risk, and reduction of liabilities. Other secondary objectives are often important. These secondary objectives may include increased environmental awareness, data collection personnel training, and cost reduction.

Scope

The scope of the audit program also must be clearly defined, preferably in writing. It should address five important dimensions: geographic location, business units, audit topics, standards, and time frame. This definition must clearly spell out which business units, facilities, and operations will be evaluated by the program. An important dimension of this issue is the geographic distribution of the facilities because regulatory requirements vary from one location to another. Another important dimension involves the business entities (e.g., joint ventures, subsidiaries, outside processors) that will be subject to audit. The definition of scope must also establish which environmental topics (e.g., air emissions, waste management, wastewater treatment) will be addressed and set forth the standards (i.e., regulations, policies, and good practices) against which the operations will be judged. Finally, the definition of scope should establish the time period to be covered by the audits. This in turn will establish a frequency for audits.

Methodology

When the objectives and scope of the audit program have been established, the company must develop a methodology for carrying out the audits. These methodologies generally fall into three categories: self-audits, centralized audits, and decentralized audits with centralized oversight. Selecting the proper methodology depends on the resources available for conducting the audits, the capabilities of the audit group, and the culture of the organization. Each of these methodologies has distinct characteristics.

Self-audits typically are driven by developing checklists and protocols that are distributed to all of the operating functions and facilities. Managers or environmental personnel at the operating facilities are required to complete the checklist by a particular deadline and establish a corrective action plan. This approach often is helpful for enhancing the awareness of environmental issues throughout the organization, but it relies solely on the diligence and judgement of the individual auditors. This approach offers the greatest potential for inconsistent interpretation of environmental standards and application of the checklists. The checklists must be updated frequently to reflect changes in requirements.

Centralized audit programs typically are administered by a full-time audit manager or by an environmental manager on a part-time basis. Teams of experienced individuals are established on a rotating basis to conduct the audits. Personnel for the teams are gathered from different parts of the organization for a period of usually less than 1 week. Because the team members have different backgrounds and different areas of expertise, it is important to train the team members to ensure good understanding of the audit's objectives and scope, auditing techniques, and the operations being reviewed.

Decentralized audit programs allow for adjustments that may be necessary to accurately review operations in different business units or in different geographic jurisdictions. Each business unit typically establishes an audit program, but to ensure communication to the highest levels of the organization and to enable directors and officers to manage environmental risks, a central group establishes an oversight program. The central audit group may establish minimum requirements for the individual audit programs and conduct internal quality assurance audits on a regular basis.

Staffing

Quality assurance audits seek to determine whether the best people in the organization are made available to conduct the environmental audits. This is one of the best ways to judge the organization's commitment to the program.

The specific staffing of an audit program depends on the objectives, scope, and methodology of the audit program as well as the organization's culture. The level of staffing must be adequate to conduct the required number of audits. The audit staff and audit teams also must possess the proper knowledge of the program's objectives, scope, and methodology, regulations, operations, and auditing technique, to conduct thorough and effective reviews.

Perhaps the most important characteristic of effective auditing personnel is an ability to work credibly with a wide range of individuals, from the boardroom to the control room. Many of the issues identified during audits require technical solutions, but their significance must be interpreted and communicated in a financial or strategic context. Experience in the line organization is one of the best ways to develop this credibility.

Many companies borrow personnel from other parts of the organization to complete the desired number of audits and to ensure that the teams have the right blend of experience and expertise. If the audit staff or audit team members are inexperienced or lack the necessary skills, training is needed to ensure the effectiveness of the program. A written curriculum and formal training are best. Otherwise, inconsistencies may develop because of differing priorities or interpretations as audits are conducted on an ongoing basis.

Organization

Environmental audit programs must reside at the proper location in the organization to exert the proper degree of influence and bring about real improvement in environmental management. In general, this means that the person responsible for the program should report to at least the vice presidential level. In the absence of a prominent position in the company, strong reporting mechanisms should exist to ensure that major issues are properly considered and addressed in a timely fashion.

Documentation of Findings

The findings of environmental audits must be properly documented. Two types of reports typically are used: descriptive reports and exception reports. Descriptive reports include descriptions of the operations along with discussions of compliance status, hazards, risks, and potential liabilities. Exception reports indicate only those situations where deviations exist from requirements or good practice.

Written reports should be prepared on a timely basis, and they should be distributed to key personnel who can commit the resources of the company to improve compliance status, reduce risks, or achieve other objectives. Adequate opportunities should be provided to review a draft report, to ensure accurate description of operations and proper characterization of risks. If the report contains recommendations, they should be flexible enough to allow facility personnel some latitude in determining how improvements will be implemented, but they should be definitive enough to ensure continuous improvement.

Regardless of the type of written report employed, the findings and recommendations should be prioritized so that resources can be properly allocated to address the most critical issues first. It is often helpful to organize the findings and recommendations into several categories, depending on their import and the level of authority needed to address them. These categories should be highlighted in the written report to enhance understanding as well as for the convenience of the reader.

The quality assurance audit also evaluates how carefully the findings are documented and managed. In preparing the audit reports, it is essential to realize that they are not generally protected from legal discovery. Reports should be prepared with the expectation that they eventually will be available to the government or a third-party litigant.

With the potential for discoverability in mind, it is also important for the company to properly manage the documents collected or prepared during an audit. No extraneous documents should be kept in the files. Copies should be held to a minimum. All documents should be properly marked (e.g., attorney-client privilege, confidential, proprietary) in accordance with the approach selected by the company.

Follow-up Mechanisms

The follow-up may be the most important aspect of an audit program. Simply stated, it is not worth conducting audits if the company is not committed to following up on findings. Liabilities can multiply, and adverse publicity can result through inaction. Companies also may increase the likelihood that they will face criminal prosecution if they lack a plan for dealing with violations detected during internal environmental audits.

To ensure progress, it is essential to clearly delegate responsibility for corrective action to an appropriate person or group. In most companies, the line organization is responsible for implementing improvements. To encourage positive action, action plans normally are prepared on a timely basis to outline the action to be taken, assign responsibility, and establish a tentative deadline for completion. Action plans then are monitored on a regular basis.

Measures of Performance

As a final way of evaluating the effectiveness of an audit program, it is useful to review the measures of performance used by the company. The measures should reflect each of the key objectives of the program. They also should ensure that the scope is consistently met, the methodology is consistently applied, and corrective actions are implemented in a timely fashion. Quantitative measures of performance can also facilitate the establishment of goals for future development of the audit program.

SUMMARY

In summary, a quality assurance audit of an environmental audit program serves much the same purpose as an annual physical or a 50,000 mile checkup on your car. Although no serious problems may be apparent, it is useful to periodically look under the hood and determine what trends have developed and what improvements can be made.

Through a quality assurance audit, management can assure itself that the environmental audit program has the resources and systems in place to identify important issues, evaluate them, and continually improve the operation of the company's facilities. With the growing emphasis of all levels of government to enforce environmental regulations, the company can save money in the form of fines and penalties. It also can avoid adverse publicity. In fact, the most effective audit programs can identify opportunities and help the company achieve its strategic goals.

5

Employment of an Environmental Auditor

Jacqueline Shields and Jill Barson Gilbert

PURPOSE OF THE AUDIT

In the 1990s, auditing has become a mature field of environmental work that is recognized as a separate discipline with identifiable practitioners. In the developmental years, auditing was performed primarily to achieve compliance and avoid litigation. Now, however, environmental audits are performed for a wide variety of reasons. For normal operating conditions, audits are performed to determine:

- The current status and/or history of compliance;
- The current status and/or history of enforcement actions;
- The current status and/or history of safety and health;
- The status of pollution prevention and waste minimization programs;
- The potential for the occurrence of an emergency situation;
- The potential for liability;
- Existing litigation and liability;
- The feasibility of property transactions;
- The status of specific systems or operations; and
- Any one or a combination of any of the above.

Audits also are performed today as an integral part of the recovery phase of emergency management. In the case of a catastrophic emergency situation in which lives have been lost or bodily injury occurs, an audit may be required by the responding regulatory agencies and/or by insurance companies. Certainly, prudent corporate management will insist that the auditing be initiated as soon as safety permits. This is a special type of audit that has unique requirements, both in content and auditor competency.

As the field of environmental auditing has grown and matured, companies and individuals have begun to specialize. They may do this by performing only certain types of audits for specific businesses and industries or for particular operations within a facility. Most consulting companies, however, continue to offer general auditing as their primary service, with specialty audits as a secondary service. Not only are environmental auditing companies growing in size, but environmental auditors are rapidly increasing in numbers. Clients are now being aggressively pursued.

Before any decision is made to hire an environmental auditor, the purpose of the audit should be clearly defined. This decision will govern all aspects of the audit, including the type of environmental audit to be conducted, the composition of the audit team, the quality control of the audit and auditors, and the delivery of the report at the end of the audit. The rationale for conducting the audit should also consider how comprehensive the audit will be, the type of report required, and how many facilities will be included.

The purpose of an audit in normal operating conditions can be determined after due deliberation by corporate staff after consultation with all affected parties. Prudent management and corporate decision makers know that it is always better to make the decision to audit rather than have the auditing decision imposed by external sources. They should, therefore, seek to arrive at the decision regarding the purpose of the audit in the most timely manner possible. Unfortunately, an emergency situation may exist and such time may not be available. In this case, the auditing decision may be made by a regulatory agency. In such a situation, the time involved in conducting the audit will be an important factor.

HIRING OPTIONS

Whatever the purpose of the audit, after the decision has been made, the question invariably arises about who will perform the audit. At this point, there are three major options to be considered:

1. Using an in-house auditing team;
2. Using an in-house auditing team in combination with internal or external consultants; or
3. Hiring a consultant (individual or company) that performs general and/or specialty audits.

The first and second options are the ones most commonly used by large businesses or industries. They have the financial means and resources required to assemble, train, and maintain an auditing team. The size of their operations necessitates a larger volume of auditing, thus maintaining an in-house auditing team may be a more cost-effective option. By having an in-house team perform the audit, larger businesses or industries can enjoy the advantage of having their audits performed by a group of experts who have an intimate knowledge of all phases of their operations. The company also may elect to have a small corporate headquarters team of auditors who travel to the company's various facilities to perform audits, assisted by individuals who work at the facility. This has the extra advantage of standardizing the auditing process throughout the company. These companies will have developed their own criteria for the composition and duties of the auditing team. Many of these criteria will be similar to those used when hiring a consultant.

The second option of using a mixed auditing team composed of in-house expertise and external consultants is the one most favored by medium- to small-size companies. This option allows the company to maintain close oversight of the auditing process, yet avail itself of the services of experts. External consultants bring a fresh perspective to the audit process and may detect areas of deficiency that have been previously overlooked in the preaudit review. Not only can it be cost-effective, but it can provide some measure of relief to a small environmental department that already may be stretched to its performance limits.

The third option is selected when the company does not have the resources or desire to maintain an in-house auditing team or auditor or when an independent auditor is required by regulations or other factors. The company may require only intermittent or limited auditing, in which case the use of an external auditor would be more cost-effective. A special situation, such as a catastrophic event, may necessitate the use of a special auditor. The company may need the services of a specialist that cannot be found in house. The timing of the audit may be such that additional outside help may be needed to meet deadlines. In the case of a compliance review, the company may feel that a third-person review is needed. Also, a company may not be fully aware of the requirements of new environmental legislation and can use an external auditor to update the compliance review.

Because of the complexity and range of the Clean Air Act (CAA) Amendments of 1990, many companies, both large and small, will find themselves in this situation. Although companies with a large environmental staff and a permanent auditing team will have the advantage of familiarity with the 1990 CAA Amendments, they may use the services of an external consultant in their compliance quality assurance program. Certain sections of the 1990 CAA Amendments, such as the Hazardous Air Pollutant (HAPs) regulations or the acid rain provisions, also may require a company to initiate new pollution control programs in which it has little experience or expertise. Major changes in permits and permitting procedures are also anticipated when the final regulations for the CAA Amendments are promulgated. Under these conditions, should the company need to quickly develop and implement new pollution prevention programs, it may be necessary to hire a consultant to assist the company's regular auditing team.

ESTABLISHING BOUNDARIES

After considering the advantages and disadvantages of each of the three options, the company will finally select the auditor or auditing team that can best achieve the purpose of the audit within the established operational boundaries. The boundaries imposed on the auditing process should be clearly defined before the selection of the auditor begins, since the auditor may not be able or willing to work within these boundaries. Given the present economic climate in the United States, the cost of performing an environmental audit, and the expenses incurred in installing new equipment or systems, the first limitation will probably be financial.

The second limitation may be the time limit set for completion of the audit. This will be very important when compliance with the provisions of the 1990 CAA Amendments is an issue. Prudent companies already have begun to audit facilities for the presence of any of the 189 listed HAPs. Present HAPs control systems may not be adequate to meet the new HAPs regulations. System modifications may have to be done as quickly as possible in order to be in compliance when the new regulations become effective. If major system modifications or large capital investments in new abatement equipment are required for compliance with the 1990 CAA Amendments, both time and cost become limiting factors. If a small environmental staff must perform the audit in addition to all of their routine work, they may not be able to perform the audit in the specified time frame. It becomes obvious that not only will limitations be circumscribed by cost and resources, but also by new environmental legislation.

CRITERIA FOR HIRING THE AUDITOR

Once the purpose of the audit and the boundaries within which the audit is to be performed have been established, the next task is that of hiring an auditor. The auditor may be an individual or a consulting company; however, in either case there are certain common criteria that can be evaluated when making this decision. As in all sound business decisions, it is wise to compare alternatives, and this can be done more effectively if criteria are set and presented in writing.

A form listing the basic criteria for evaluating prospective auditors is given in Table 5-1. This form can be used not only as a checklist, but also as a summary of the information collected during the interview process. Simple checkmarks in each category can be used to indicate that an item has been discussed. The checkmark also can be accompanied by a reference number that indicates that detailed information can be found. When a more detailed version is required, information and data can be entered directly on the form. Since this form can be computerized, modifications in the size of the columns can be made easily to allow sufficient space for information and comments to be included. The company may elect to take each item in sequential order or may examine several

TABLE 5-1 Criteria for Hiring an Auditor

CRITERIA	AUDITOR A	AUDITOR B	AUDITOR C
Existing Conditions			
Purpose of Audit			
Boundaries			
Qualifications			
Type of Company			
Numer of team members			
Additional personnel resources			
Non-personnel resources			
Area of expenise			
Certification			
Other credentials			
Technical expenise			
International experience			
Performance			
Compatibility of corporate philosophy			
Willingness to work with company staff and management			
Access to external data & information sources			
Security systems employed			
Reporting capabilities			
Estimated cost of audit			
Estimated time to perform audit			
Guarantees of performance			
References			
Client satisfaction			
Relationship with agencies			
On-time, cost overruns			
Reliability			
Availability			
Conclusion of contract			
Comments			
Weighted-rating of auditor			

categories simultaneously. In the interest of time and efficiency, the latter is recommended, since the verification of references should not impede the progress of checking either the auditor's qualifications and resources.

The form also includes separate columns headed by the name of each auditor being considered for employment. This provides a rapid visual means of comparison and also quickly identifies those areas where differences may occur. Such a method of comparison also minimizes confusion that may arise when unequal or different items are mistakenly compared on the same basis. This form also can be used in an auditing quality assurance program, particularly when regulatory agency auditors have been involved in the auditing process. Although regulatory agency auditors usually will not be employed by the company, it may be prudent to perform an evaluation of their qualifications and performance using the same criteria that were applied to the company's auditor. Given the liabilities and penalties that can be incurred in the current regulatory climate, it is not unreasonable to document such an evaluation.

CRITERIA CATEGORIES

As shown in Table 5-1, the criteria for hiring an auditor have been divided into four major categories: existing conditions, qualifications, performance, and references. Both the auditor and the company should have a clear understanding of the purpose of the audit and the boundaries within which it will be conducted. A written statement of both of these should be included, to serve as a constant reminder to both parties during the course of the criteria evaluation and hiring process.

Qualifications

The second category of "qualifications" properly identifies the type of auditor—that is, an individual, a consulting company, a company audit team, or a regulatory agency auditor. The numbers of individuals that will compose the audit team should be listed. This includes company employees not usually associated with the auditing process, and special consultants, no matter how brief their involvement in the audit. A number can be given on the audit, with an attached listing of all individuals who will participate, or both can be listed on the form.

The auditor's area of expertise should be identified next. As previously stated, while many auditing consulting companies conduct routine audits of facilities, they also maintain teams or individuals who can perform special types of audits for unique systems, special legislative requirements, or certain types of facility operations. Notations should be made of certifications or other credentials that may be relevant to the particular audit, such as the Professional Engineer licensure or Certified Industrial Hygienist.

It also should be remembered that yesterday's expertise may no longer be adequate to meet today's regulatory requirements. An example of this is the changes imposed by the new regulations of the 1990 CAA Amendments in the auditing of air emissions and abatement systems. Many environmental auditors have included the auditing of air emissions and air quality control operations in their listing of services. It should be remembered, however, that, while this experience is helpful, it will not compensate for the lack of preparation to meet the complexity and range of requirements contained in this new legislation. The early identification of expertise can serve as an upfront screening device and can save time and money in the hiring process.

Also, the international experience of the auditor should be documented and verified. This includes the identification of existing resources of personnel and equipment in other countries and familiarity with the political situations, regulatory requirements, customs, and languages of those countries.

Performance

As early as possible in the evaluation process, a determination of the compatibility of the auditor with corporate philosophy should be made. It is assumed generally that the facility manager and the corporation share the same ethics and operational philosophies. This, however, may not be the case and will rapidly become apparent when the compatibility of the auditor's and company's beliefs and ethics are rigorously examined. No company wants to see years of hard work in establishing a positive regulatory agency relationship destroyed in one afternoon by an auditor with an aggressive, adversarial attitude. The auditor must be willing and able to work with not only the client's staff and management, but also with regulatory agencies, to promote the client's best interests. The auditor's ability to communicate on different levels in each of these situations is most important.

The technical expertise should be evaluated not only in terms of technical knowledge and experience, but also by capabilities in field testing and monitoring and analytical laboratory work. Particular attention should be given to quality assurance programs for each phase of work.

The access to external data bases and information sources should be considered an integral part of these resources. The type and form of data base used to record and store information and the data base collected during the audit should be identified. Particular attention should be paid to the security of the system. A notation should be made of the security systems, and precautions should be used to insure confidentiality of the audit results and to protect company patents and secrets during the auditing process. Signed agreements to protect trade secrets or confidentiality agreements may be needed.

Audit reports may be verbal, written, or a combination of these. They also may contain facts and findings only or may include recommendations and action plans. Whatever the type of report required, the auditor's reporting capabilities should be evaluated in terms of the ability to produce a professional-quality final report, develop visual graphic presentations, and translate the report into languages other than English. If possible, examples of reports for previous clients should be reviewed to determine the expertise and capability of the auditor in reporting results.

Inquiries concerning the cost of performing the audit can be done in stages. The cost of the work on the audit may be charged as hourly rates only or as a total project cost. It is not unusual to first obtain a gross estimate of the cost of conducting the audit for cost comparison purposes and also for later contract negotiations. Depending on the extent or level of the audit, prices range from around as low as $100 for the environmental endorsement of a homeowner's real estate transaction, to $100,000 or more for a comprehensive audit of a large industrial site that has been in operation for over 20 years. An audit for medium to small facilities should cost between $5,000 and $20,000. Again, much will depend upon the thoroughness of the audit required by the company and the previous attention to good environmental practices and compliance.

After the cost estimates have been accepted and the decision has been made to hire a particular auditor, it is customary to request a detailed, accurate budget along with the statement of deliverables and work schedule guarantees. At this point, the company or auditor may wish to conduct further negotiations to resolve questions of cost, deliverables, guarantees, or a combination of all three. It would be wise to have legal counsel available during the negotiations and for the review of proposed contracts.

Given adequate information regarding the scope of the work required, the auditor should be able to provide an estimate of the time required to perform the audit and submit the final report. The inability or unwillingness on the part of an auditor to provide such an estimate may be an early indicator that further negotiations would not be productive. The time required to complete the audit and final report may be of utmost importance, particularly if a major compliance problem exists or an audit is required by a regulatory agency. Given the massive changes in operations that are anticipated under the 1990 CAA Amendments, many companies are beginning to move swiftly to complete audits to determine what changes must be initiated to achieve compliance. As with cost, the exact date for completing the audit and submitting the final report can be included in the contract negotiations.

References

This category is the "Achilles Heel" in the entire process of hiring an auditor because it is often the one that receives least attention. References should be verified at least by a telephone call. An effort should be made to obtain a letter of reference; however, given the problems of liability and litigation, it is unlikely that many companies will be willing to provide this type of reference. The results of the inquiries should be noted on the criteria form (Table 5-1). Both the professional reputation of the reference and the length of employment of the auditor should be noted. The information obtained from formal as well as informal inquiries should be recorded.

Client satisfaction is first on the list in this category. Specifics should be requested! Questions should be asked about the auditor's technical competence and ability to handle sensitive situations, particularly those involving regulatory agencies. This also provides the opportunity to discover those areas where performance did not meet expectations. The ultimate question is, "Would you hire this auditor again?"

References also should be asked to provide information on cost overruns, timely completion of all work, ability to give quick turnaround on the project, and the reliability of the auditor. The availability of the auditor after the audit has been completed may not be a prime consideration in selecting an auditor, but references should be asked to comment about this. This may become an importance factor if problems associated with the audit arise after the audit has been completed. The contribution of the auditor to these problems or the assistance of the auditor in resolving these problems should receive particular attention.

FINAL SELECTION OF THE AUDITOR

When all references have been contacted, all categories checked, all interviews conducted, and all costs considered, it is time to make the final selection of an auditor. The choice may be obvious, in which case hiring the auditor will simply be a matter of signing the final contracts for employment.

It may, however, be a difficult choice between two auditors who have comparable qualifications. In this case, the hiring may be made on the basis of personal rapport established during the interview process or of past professional associations. If this is not a satisfactory solution, the problem may be resolved by using a weighted rating approach. This involves assigning a weight of importance to each of the categories. The auditor is then rated in each of the categories. A summation of the weighted value multiplied by the rating value is done. The auditor with the highest score then should be selected.

Selecting an environmental auditor is not only a serious endeavor, but it is also a very personal matter, from an individual as well as a corporate perspective. This is inevitable to some degree, given that the auditor will be permitted not only to observe all operations at a facility, but also have access to proprietary information. Throughout this entire process, the auditor will be expected to render an objective evaluation of the successes and failures of operations, as well as the contributions of corporate and management to each. Given these circumstances, it is natural to have strong tendencies to choose the most sympathetic auditor. Much of this can be minimized by the application of a systematic method of evaluating the qualifications and past performance of an auditor. This chapter has presented a system that can be modified to meet the particular needs of a company. The use of a systematic approach to evaluating auditor qualifications and capabilities, combined with the collective decision making of both corporate and management, does not guarantee a perfect choice of an environmental auditor. It does, however, maximize the probability of success in the selection of an auditor that best matches the short- and long-term needs of the company.

References

Clean Air Act Amendments of 1990, PL 101-549, November 15, 1990.

Guida, J.F. A practical look at environmental audits. *Journal of the Air Pollution Control Association*, Vol. 32, No. 5, pp. 568–573.

Gulledge, W.P. 1989 Registration and certification of environmental professionals: the insurance perspective. *Working Papers*, Vol. III, No. 1, Spring, p. 1.

Priznar, F.J. 1990. Environmental auditing organizations. *Working Papers*, Vol. 4, No. 3, Fall, p. 11.

III

Specific Situations

6

Air Quality Data Sources for Audits

William P. Gulledge

INTRODUCTION

In auditing air emissions programs, the following information is required:

1. Types of data available;
2. Sources for that data; and
3. Some analysis to use the data that are acquired.

The audit can be structured such that only compliance parameters, such as emission limitations or new source permit conditions and monitoring, are assessed, or additional air emissions programs (pollution prevention or air toxics reduction) also may be reviewed. Compliance data are centered on criteria and precursor pollutants for compliance with National Ambient Air Quality Standards (NAAQS).

Non-criteria pollutants must also be considered; hazardous air pollutants (number expanded to 189 Compounds under the 1990 (Clean Air Act) CAA Amendments) and fugitive emissions are examples of these pollutants. The audit may focus on the data used to establish compliance, or it may be targeted to determining the effectiveness of the program used to collect data. Alternatively, the management system used to achieve compliance or post-compliance reductions can be assessed.

AIR QUALITY DATA SOURCES

Data on emissions that contribute to attainment or non-attainment with NAAQS have been collected since the initial amendment of the CAA in

1970. Ambient air quality monitoring networks exist to measure concentrations of sulfur dioxide, nitrogen oxides, carbon monoxide, particulates, and ozone. An acid precipitation monitoring network has also been added. These data are collected by the Environmental Protection Agency (EPA) and published annually in an EPA report on the state of the environment. Results can be used to monitor progress in meeting or exceeding NAAQS.

NAAQS data obtained for the 1980s indicate that concentrations for all ambient air quality pollutants except ozone have been reduced during the 1970s and 1980s. Sulfur dioxide has experienced the most significant reduction. Ozone concentrations have consistently been high in hot weather months. Although not universally true, the number of days for ozone exceedances is usually greater during the summer months.

Computerized information systems usually offer the most convenient access to air quality data. Some computer records also are updated more frequently than published EPA records. There are many types of computer data bases that relate to air emissions and other environmental parameters. There are two main functions for databases:

1. Information storage—Many systems are used to store monitoring or compliance information. Computations or statistics cannot be performed within these storage databases. If this is the only type of data available for a location being audited, the data should be reviewed for compliance or other audit purposes.
2. Calculational data storage—Information can be stored on spreadsheets in which the user can perform statistics and other computations. For air emissions information, mathematics can be conducted to average samples and to determine daily, monthly, and annual averages for compliance.

There are over 300 governmental data bases that relate to environmental impact information. Some of these sources provide health effects data for specific environmental media, while others focus on source emission measurements. Computer software has been developed to audit air emissions compliance parameters.

The National Emissions Data System (NEDS) is maintained by the EPA and contains emissions data for sources emitting more than 100 tons per year of criteria air pollutants. Data on volatile organic compounds (VOCs) and particulates is included. The EPA's Toxic Release Inventory System contains specific source data on emissions of toxic and hazardous air pollutants.

Some computer record searches primarily are used by parties conducting Phase I environmental site assessments. In most instances, air emissions are not the major focus of these assessments, and the data search companies do not concentrate on air quality data. Most of the services obtain toxic release inventory information that may have been filed for a specific site. Data on criteria pollutants are normally not available from these sources.

The following companies specialize in computerized information record searches for property transactions. State-of-the-art information obtained and analyzed from these services is changing rapidly. Newer geographic information services are being integrated with government data base systems to produce more useful, specific-location environmental impact data. Many of these companies are currently going through a consolidation and acquisition phase, so this listing is not complete or necessarily accurate. Leading companies are:

- The Environmental Risk Imaging and Information Center;
- VISTA/Environmental Audit; and
- Environmental Data Resources/Toxicheck.

Many environmental consulting firms also provide software to assist companies with regulatory compliance. The 1991–1992 *Directory and Resource Book* of the Air and Waste Management Association lists 24 companies as suppliers of software for toxic chemical tracking and compliance. The extent to which these companies also provide compliance tracking information for criteria pollutants is unknown.

Air emission data are also available as outputs from air quality models. These models may be dispersion models for point source emissions, ambient air quality forecasting, or toxic air pollutant models for sudden, accidental releases. Like measured air quality data, care should be taken to make sure that air pollutant information provided from models is screened for accuracy and quality control.

Emission factors for volatile organic compound releases can be obtained from the EPA's *AP-42 Handbook.* Care should be taken when using the emission factors, since the factors were developed from specific industry types. These factors may not be applicable to the industry in which the auditor is working. The Air and Waste Management Association recently published an updated version of their *Air Pollution Engineering Manual* that includes emissions information for a variety of air pollution sources.

Environmental agency records are another source of air emissions information. Records may be available on the federal, state, or local level. Federal information is available from EPA regional offices. This information includes ambient air quality monitoring and permit information for those facilities holding a New Source Performance Standard (NSPS) permit.

If basic information on air emission health and environmental effects or emission control technology is needed, many EPA documents have been published containing this information. These documents include:

1. Source assessment documents;
2. Background documents for control technology guidelines; and
3. Background documents for New Source Performance Standards for Hazardous Air Pollutants (NESHAPS).

Health effects documents are available from the EPA's Office of Air Quality Planning and Standards and the Office of Research and Development. Emissions monitoring data submitted to maintain or determine compliance with permits may be available for selected facilities.

Both measured and estimated (modeled) data on toxic air emissions are available from SARA Section 313 inventory submissions (Form R). Section 313 data include process and fugitive emission releases. Most of the major industry types operating in the United States are required to submit Form R information for each facility meeting the reporting threshold requirements. Form R summary information is becoming a more important indicator of air emission releases for non-permitted emissions.

Some states may collect additional air quality data. This is especially true for states in industrial areas (Texas, New Jersey). Data may be available for criteria pollutants (SO_2, NO_x, CO, and PM_{10}). Toxic air pollutant information also may be available if a state or local metropolitan area has collected toxics information. Data from regional monitoring networks, such as the Houston regional ozone non-attainment area, are also helpful in establishing baselines and monitoring trends.

Most requests for government agency ambient air quality standards information can be easily handled, since most of this information is readily available. Rarely do Freedom of Information Act (FOIA) requests need to be filed for these data. FOIA requests may be necessary for obtaining specific facility permit compliance information. All data requests should be directed to the air quality section of a state's environmental regulations department.

Other sources for air quality data include research and private organizations. Many of the private data collection efforts are for specific purposes. For instance, the Electric Power Research Institute maintains an acid deposition monitoring system and is sponsoring research on utility sources of toxic air pollutants. The Chemical Manufacturers Association has sponsored data collection efforts related to fugitive volatile organic compound emissions. The American Petroleum Institute has obtained air emissions data from refinery operations. Data from private organizations may or may not be available for public use.

Health effects information for exposure to air pollutants is available from the Center for Disease Control (especially for indoor air and tobacco smoke), the EPA, and the U.S. Department of Labor, Occupational Safety and Health Administration (OSHA). The OSHA information includes the basis for establishing occupational threshold limit values for certain air pollutants. Health effects data include clinical studies as well as epidemiological information on exposure to criteria pollutant precursors and certain toxic air pollutants. Information on radon health effects, monitoring, and remediation technology is available from the EPA and the Department of Housing and Urban Development.

TYPES OF DATA TO BE AUDITED

Auditing for compliance with air quality regulations or other criteria entails a review of records related to emissions, permits, control technology effectiveness, and management and monitoring practices. Data can pertain to particulate matter, criteria pollutants, and non-criteria pollutants. A checklist for reviewing information is normally used as part of the audit procedure. Most companies conducting environmental audits have developed their own audit protocols and checklists.

Air quality data can originate from measurements of ambient air or source monitoring. Data can also be generated from air dispersion modeling. The quality of the information used for compliance certification or other purposes should be carefully reviewed as part of the audit procedure—especially information appearing as outliers of statistical analysis for a given monitoring source.

Particulate matter is an important compliance parameter. Total particulates are used for determining compliance with site permit conditions. PM_{10} is used to compare emission limitations with federal particulate standards and the state permit for a given facility. Particle sizing is also important for determining proper air pollution control equipment. Poor removal efficiency or long downtimes that are attributed to maintenance problems with emissions control equipment, such as baghouses, electrostatic precipitators, and scrubbers, should be investigated in the audit.

Emission records, such as monitoring data and control technology performance, should be reviewed for criteria pollutants (SO_2, NO_x, and CO). These records should be checked for compliance with federal standards and state permit conditions. Any problems encountered should be noted in the audit notes or working papers.

Non-criteria pollutants include air toxics and fugitive emissions. Air toxics originate from manufacturing processes, motor vehicles, waste incineration, and manufactured products (indoor air emissions). Emissions from the first three sources should be checked against any standards that may be applicable for the specific facility. The manufacturer of a given product may have also established guidelines for indoor air emissions. OSHA guidelines (permissible exposure levels) should be compared to indoor air concentrations.

Form R data from the SARA Section 313, inventories can be used to monitor progress in achieving waste minimization goals. This measurement can be made from a total toxics reduction, a toxic air emissions reduction, or a pollutant-specific reduction. Since Form R information is publicly available, many companies are tracking waste minimization progress.

Fugitive emissions and secondary emissions are produced from manufacturing process leaks or secondary releases into the environment. Flanges and valves associated with petroleum refining, petrochemical processing, or organic chemical manufacturing are a major source of fugitive emissions. Benzene fugitive emissions are regulated as a NESHAP.

Secondary emissions occur as air emissions from wastewater and hazardous waste disposal facilities or as VOC air emissions from groundwater remediation projects. Secondary emission standards may exist at the state or local level. Much of the information used to determine secondary and fugitive emission releases is generated from modeling rather than measurements at the source. Both techniques are frequently used. Modeling and monitoring data should be carefully checked if compliance is an issue.

Indoor air data are used in audits designed to monitor ambient air concentrations or emissions from specific products. Carpets, plywood, and assorted plastic materials are all sources of relatively minor toxic air releases. However, for sensitive individuals, the emission from these sources can cause significant discomfort. Many manufacturers monitor the emissions produced by new products in the microgram per liter range.

If time permits, all existing air emissions data should be reviewed during the audit. If the data are too extensive to permit a comprehensive review, a representative sample should be selected. Exceptions, comments, or problems discovered should be included in the audit process and communicated either verbally or in writing to site management.

Appendix 6-A

Air Pollution Audit Materials*

*From *Environmental Auditing Workbook*, prepared by Arthur D. Little, Inc, for the Edison Electric Institute, 1983.

NOTICE: This publication is designed to provide accurate and authoritative information with regard to the subject matter covered. However, it is provided with the understanding that the contents reflect Edison Electric Institute's and Arthur D. Little's best judgment in light of the information available at the time of preparation. Neither Edision Electric Institute nor Arthur D. Little make any warranty or assume any liability with respect to the use of, or damages resulting from use of, any information contained in this publication. If specific compliance advice or assistance is required, the services of a competent professional should be sought.

Publication of this Workbook does not signify that the contents necessarily reflect the views and policies of Edision Electric Institute.

AIR POLLUTION AUDIT PROTOCOL

> This audit protocol is intended as a guide for designing and conducting facility audits. It may require additions, revisions, or other modifications in order to meet the needs of your particular audit objectives, industrial setting, or other special circumstances.

Facility Name: ____________________ Date(s) of Audit: ________________

Team Members Participating in Air Pollution Audit: ____

Protocol Prepared By: __________________ Protocol Reviewed By: __________________

	AUDITOR(S)/ COMMENTS	WORKING PAPER REFERENCE (List Page Numbers)
1. Using data supplied by the plant or provided by the audit team leader, review the following:		
a. Production process flow diagrams and descriptions including control devices;		
b. Facility layout, including location of all stationary sources;		
c. Emission inventory;		
d. Regulations—federal, state, and local (including attainment/non-attainment status for each pollutant);		
e. Any outstanding court orders: variances, compliance orders, administrative orders, etc.;		
f. Applicable corporate policies, procedures, and standards;		
g. Facility policies and procedures;		
h. Operating manuals for air pollution control systems;		

	AUDITOR(S)/ COMMENTS	WORKING PAPER REFERENCE (List Page Numbers)
i. Air pollution alert and emergency plans;		
j. Operating and construction/modification applications and permits.		
2. Review the Air Pollution Questionnaire completed during the advance visit. Familiarize yourself with plant responses. (If questionnaire was not administered in advance, complete Air Pollution Questionnaire.) Make inquiries as desired to increase your understanding of the plant's responses.		
3. At some time during the audit, tour the plant following production processes and facility layout of stationary sources:		
a. Using the process flow diagrams and plant layout, locate and note all points of continuous and periodic emissions;		
b. Where possible, list pollutants in emissions;		
c. Record location of all emission control and monitoring facilities;		
d. Note any likely sources of emissions not included above (i.e., fugitive dust, process losses, etc.);		
e. Note location and type of activity of the plant's neighbors (i.e., industrial, residential, institutional, etc.).		
4. Document in flow chart or narrative form, your understanding of the following:		
a. Air pollution control systems and procedures;		
b. Continuous monitoring;		

	AUDITOR(S)/ COMMENTS	WORKING PAPER REFERENCE (List Page Numbers)
c. Periodic monitoring and sampling programs and procedures; d. Procedures for recordkeeping and reporting of compliance; e. Testing and analytical procedures; f. Information handling and documentation procedures for (b) through (e) above. 5. Prepare a list of all active, pending, and inactive regulatory permits, operating certificates and registrations for airborne emissions (during the period under review). Include notation of agency, facility, type of permit, effective date, expiration date. a. Confirm that all identified emission sources are covered by permits; b. Review for completeness the applications for registration, certification, operation, and construction; c. Determine that applications are accurate and signed by appropriate company personnel; d. Check compliance with PSD nonattainment rules. 6. Develop a schedule of capital expenditures made during the period under review, and confirm that appropriate permits were obtained for all situations in which they were required. 7. Using the emission sources developed in Step 3, and the permit data developed in Step 5: a. Determine that all applicable standards, limitations, and compliance schedules are being met for regulated emissions from all sources;		

	AUDITOR(S)/ COMMENTS	WORKING PAPER REFERENCE (List Page Numbers)
b. Determine through inquiry, observation, and review of available documents that the plant is in compliance with applicable opacity standards;		
c. Determine that emission and fuel monitoring and test methods and procedures followed are in accordance with applicable regulations;		
d. Review supporting documentation noting that all instrumentation, analytical techniques, calculations, records, and sampling locations and methods are in accordance with state and local regulations.		
8. For those plants with requirements under CFR 40, Part 60.7 (new source performance standards), complete the following:		
a. For the period under review, examine regulatory required monitoring reports submitted in compliance with CFR 40, Part 60.7 (new source performance standards);		
b. Review emission and ambient air monitoring records for completeness in accordance with CFR 40, Part 60.7 (d) (new source performance standards). The review should include at least:		
(1) Determination that performance tests were in accordance with CFR 40, Part 60.8;		
(2) Review of maintenance department records for calibration, reliability checks, and adjustments of monitoring devices;		

	AUDITOR(S)/ COMMENTS	WORKING PAPER REFERENCE (List Page Numbers)
(3) Investigation and documentation of any unusual items noted during the file review. c. Review plant's monitoring systems noting compliance with CFR 40, Part 60.13. The review should include at least the following: (1) Determine that monitoring systems and devices were installed and operational *prior* to conducting performance tests; (2) Confirm that performance specifications noted in Appendix B of CFR 40, Part 60 are met; (3) Determine that the manufacturer's written requirements or recommendations for checking the operation or calibration of the devices were carried out; (4) Document the adequacy of any alternative monitoring requirements used in the absence of a continuous monitoring system. 9. For the latest National Emission Standard for Hazardous Air Pollutants (NESHAP), determine that permit or standard limitation is being met. a. Where applicable, establish that approved testing and reporting procedures are being complied with, especially for: (1) Asbestos; (2) Beryllium; (3) Mercury; (4) Vinyl Chloride;		

	AUDITOR(S)/ COMMENTS	WORKING PAPER REFERENCE (List Page Numbers)
(5) Lead; (6) Benzene; (7) Acrylonitrile.		
10. Review hazardous air pollutant source documents applicable to construction, modification, or operation of stationary sources. Check the following: a. Timeliness with which the information was filed; b. Completeness of the information forwarded to the regulators; c. Approvals and/or subsequent correspondence from regulators (i.e., waiver of compliance).		
11. Review the plant's air pollution alert and emergency plan and note any significant departures from regulations and determine that they are up to date.		
12. Review a listing of public nuisance complaints (e.g., odors, dust, etc.) for period under review.		
13. Prepare a schedule of requirements under any outstanding court orders (see Step 1) and verify that all requirements are being met.		
14. Confirm that the sulfur content of fuel oil used for fuel-burning equipment is less than that prescribed in applicable regulations.		
15. Determine that the fuel dispensed for motor vehicles is in compliance with applicable federal regulations.		

	AUDITOR(S)/ COMMENTS	WORKING PAPER REFERENCE (List Page Numbers)
16. Review questionable items with responsible plant personnel. Note explanations and disposition of such items.		
17. Prepare a detailed list of audit findings.		

AIR POLLUTION QUESTIONNAIRE

> This audit questionnaire is intended as a guide for designing and conducting facility audits. It may require additions, revisions, or other modifications in order to meet the needs of your particular audit objectives, industrial setting, or other special circumstances.

	YES	NO	N/A
1. Have all emission sources been identified?	____	____	____
2. Are permits in effect for all required emission sources?	____	____	____
3. Has there been any construction and/or modification of the stationary sources within the time limits of the permits or the prior audit?	____	____	____
4. Were regulatory agencies notified of any construction and/or modification changes?	____	____	____
5. Does the plant maintain records indicating the occurrence and duration of any malfunctions during start-up, operation, or shutdown?	____	____	____
6. Is a written report of excess emissions submitted to the appropriate regulatory agency as required?	____	____	____
7. Was the plant required to install any continuous monitoring systems?	____	____	____
8. Has the plant been required to conduct any performance tests of air pollution control systems?	____	____	____
a. If yes, was the test conducted under conditions specified by the regulatory agency?	____	____	____
b. Was the regulatory agency notified 30 days prior to the test?	____	____	____

	YES	NO	N/A
9. Is the plant located in an area requiring prevention of significant degradation?	____	____	____
a. Are the allowable incremental changes being utilized at acceptable rates for your operations?	____	____	____
10. Is the plant required to comply with a specified opacity standard?	____	____	____
11. Is the plant required to comply with a specified sulfur standard?	____	____	____
12. Are there any sources of visible emissions that might be concealed by plant construction?	____	____	____
13. Are permanent files maintained at the plant for:			
a. Emission monitoring?	____	____	____
b. Ambient air monitoring?	____	____	____
c. Calibration checks of monitoring devices?	____	____	____
d. Measurements for determining performance?	____	____	____
e. Records of maintenance of monitoring devices?	____	____	____
f. For at least two years following the dates of execution of calibration, monitoring, etc.?	____	____	____
14. Are any of the following present in significant amounts in plant emissions:			
a. Asbestos?	____	____	____
b. Beryllium?	____	____	____
c. Mercury?	____	____	____
d. Vinyl chloride?	____	____	____
e. Benzene?	____	____	____
f. Trichloroethylene?	____	____	____
15. Have procedures been established consistent with regulations for the proper handling of asbestos during demolition operations?	____	____	____
16. In the case of construction and/or modification, has the appropriate regulatory group approved the plant's application for air emission permits?	____	____	____

	YES	NO	N/A
17. Has there been any change in air emission sources after the effective data of standards outlined in CFR 40, Part 61, National Standards for Hazardous Air Pollutants, that has required notification be submitted to regulators?			
a. If so, were submittals in compliance with established dates?			
18. Does the plant operate under any waivers of compliance from a regulatory agency?			
19. Are all emission testing and compliance monitoring in accordance with CFR 40, Part 61, National Standards for Hazardous Air Pollutants, and/or permit specifications?			
20. Has the plant had any emissions in excess of permit limits that required reporting to the applicable regulatory agency?			
21. Is the plant in compliance with emission standards for the following at all emission sources:			
a. Suspended particulates?			
b. Sulfur dioxide?			
c. Carbon monoxide?			
d. Photochemical oxidants?			
e. Hydrocarbons?			
f. Nitrogen dioxide?			
g. Hydrogen sulfide?			
22. Are there any other substances for which permit limits have been established for this plant?			
23. Have reportable excursions on emission limits occurred more than:			
a. 6 times per quarter?			
b. 12 times per quarter?			
c. >12 times per quarter?			

	YES	NO	N/A
24. Does the plant have an up-to-date air pollution and emergency plan?	____	____	____
25. Are the plant emissions and the plant location likely to contribute to interstate or international air pollution control problems?	____	____	____
26. Does the plant have odorous emissions that result in complaints from the general public?	____	____	____
a. If yes, are records kept of the nature and source of the complaints?	____	____	____
b. Are these complaints routinely investigated?	____	____	____
27. Do fugitive dust emissions exit at the plant?	____	____	____
a. Are fugitive dust emissions from the plant a problem?	____	____	____
28. Have all permit applications been properly prepared and certified by a licensed professional engineer in the state where the plant is located?	____	____	____
29. Are all meetings with, visits by, or understandings with regulatory agencies documented (i.e., approvals for variances, future actions, etc.)?	____	____	____

Audit team member(s) completing questionnaires: ____________________

____

Location personnel providing responses: ____________________________

____

7

Monitoring

William P. Gulledge

Air monitoring data have been consistently obtained since the passage of the CAA Amendments in 1970. Criteria pollutant monitoring data were obtained throughout the 1970s and 1980s and continue to be available on at least a monthly basis. Monitoring data can be in several forms. Data can be obtained from:

1. Ambient air quality monitoring networks;
2. Point source emissions monitoring;
3. Perimeter monitors;
4. Indoor air monitors;
5. Occupational health monitors; and
6. Monitoring programs that have been developed for a special purpose or for specific compounds.

More information on toxic air pollutants was obtained beginning in the mid-1980s.

MONITORING

Regional monitoring networks were established as part of a national program to monitor compliance with ambient air quality standards. Many of the monitoring stations within these networks existed prior to 1970. All of the networks have been expanded, and ambient air quality monitoring technology has vastly improved since 1970. Resulting data are more complete and of better quality. Monitoring programs exist for ozone, carbon monoxide, nitrogen oxides, particulates, and hydrocarbons.

Ambient air concentrations of criteria pollutants are tracked through regional monitoring networks. For each criteria pollutant, trends can be assessed monthly, seasonally, and annually, as far back as the data are available. However, quality assurance and quality control for monitoring networks must be considered when reviewing historical air quality information. Care should be taken when reviewing sampling methodology and frequency and averaging times. Seasonal data are especially important for ozone, where concentrations are much higher during the summer months or during stagnant air periods.

Toxic air pollutants may be monitored in ambient air, at point sources such as stacks and vents, and for fugitive emissions from process valves and flanges. Air toxics may also be monitored indoors for occupational or residential health concerns. Urban air toxics monitoring networks have been established for some U.S. metropolitan areas, and data are available for several hundred toxic air pollutants. However, the information that is available is considered to be very preliminary in content. More data are needed to establish long-term trends for air toxics.

Most of the urban air toxics monitoring networks have been established for a specific purpose, such as the EPA's toxics integration projects. Studies for this program have been conducted in New Jersey, North Carolina, Colorado, and Texas. Typically, both indoor and ambient air toxics data are available from the toxics integration or integrated environmental management programs. Data from the monitoring programs indicate that indoor concentrations of air toxics equal or frequently exceed ambient air concentrations.

Ozone depletion has become a key issue in the management of Air emissions. Satellite images or concentration data obtained from high-altitude aircraft are being used to characterize ozone trends above the polar caps. This high-technology monitoring network is capable of detecting seasonal, monthly, or even daily trends. Monitoring for ozone depletion is expected to increase in future years. The manufacture of highly reactive fluorocarbons is scheduled to be phased out by the late 1990s. Auditing the phase-out implementation and resulting effects will be new tasks for the late 1990s.

Other forms of monitoring also may be used for toxic air emissions. Areas surrounding organic chemical manufacturing or petroleum refining facilities may be equipped with community warning systems for sudden, accidental releases of process emissions. Industrial locations may be equipped with indoor occupational air monitors. These monitors are designed to sound an alarm when the ambient concentration of the selected toxic pollutant exceeds a threshold value.

AUDITING OF AIR MONITORING RESULTS

Air monitoring data are used to:

- Establish compliance with air quality standards and permit limitations;
- Measure progress for air toxics reductions; and
- Assess performance in meeting other guidelines or policies.

The techniques for auditing air monitoring results are similar to the techniques used to audit other media. An audit protocol is developed, and, in many cases, an audit questionnaire is used to obtain information for the audit. A typical audit protocol and questionnaire, developed since the 1977 amendments but prior to the passage of the 1990 amendments to the Clean Air Act, are included in the Appendix of Chapter 6. The protocol and questionnaire are offered as examples and should be carefully reviewed and updated prior to use.

Monitoring data are reviewed in order to establish baselines for emission control equipment efficiency. This information can be used to establish environmental management status as it pertains to air emission issues. If toxic air emissions are being monitored, an assessment of the monitoring program should include:

- Adequacy of monitoring and modeling information—Is monitoring information available for all regulated pollutants that require monitoring? Is the modeling methodology current?
- Completeness of emission inventory for process, fugitive, and secondary emissions—Are significant air emission sources missing? Have all significant pollutants been included in the inventory?
- Review of assumptions used to estimate all emission sources—Can assumptions be justified on a technical basis? What data are available to support assumptions?
- Quality control review of emission monitors—Are monitors operating within accepted parameters? If monitoring includes an alarm system, have alarms been tested?

Several basic issues should be specifically reviewed for air monitoring data. These issues include:

1. Is the monitoring equipment operated on a continuous, hourly, daily, or other basis? The frequency of measurement and/or averaging time used for monitoring and compliance should be documented.
2. How often is monitoring equipment serviced? Is routine maintenance performed? If monitoring equipment is used to detect accidental releases, are the sensors regularly tested? Have emergency drills been conducted for accidental releases?

3. Are data obtained from monitoring networks subject to statistical review? Are outlier data or measurements taken from process upsets or other malfunctions discarded prior to statistical review? What statistical methodology is used to assess monitoring information?

Reviewing air monitoring information is an important component of compliance audits. The auditor should review all permits, monitoring information required under applicable permits, available data on pollution control efficiency, and housekeeping and maintenance procedures to control air emissions. The organization's management program to control air emissions should also be reviewed to determine effectiveness. Questionnaires and checklists, such as the one in Appendix 6-A, will aid the auditor.

8

Baselines

Jacqueline Shields

While the environmental compliance audit is generally conducted using a form specific for the facility being audited, it is also conducted within a framework of baselines. Although all items in the audit are measured against certain baselines, neither the auditor or audited entity may have a clear understanding of what constitutes a baseline. Simply stated, a "baseline" is the standard or level against which all evaluations are made. A baseline exists as an standard or specification and may take many forms, both quantitative and qualitative in nature.

TYPES OF BASELINES

Baselines may be numerical or descriptive in nature. Numerical baselines consist of a specific number or a set of numbers that form the boundaries of the permissible range of operations or the acceptable status of the environment. Emission limits set forth in permits and National Ambient Air Quality Standards (NAAQS) are prime examples of numerical baselines. The standards set forth in numerical terms in all environmental rules and regulations form the baseline for permitted discharges and emissions. Inventories of emissions and discharges, flora and fauna, and health effects also constitute baselines in that they quantify and qualify existing conditions.

A baseline that is descriptive in nature will specify what must be done, how it is done, where it will be done, and/or who will perform the activity. It may also define the scope of a particular activity. Examples include identifying environmentally sensitive areas or stating the requirements by some states for individuals with certain qualifications to conduct an environmental audit.

Sometimes these baselines are clearly stated. An example of this would be the limits set forth in the facility permit. In such cases, the inspector or auditor judges performance and appearance against the stated standards or specifications. The auditor will find the word "baseline" repeatedly stated and defined in the various sections of the 1990 Clean Air Act (CAA) Amendments (Appendix 8-A).

Other baselines may not be so clearly defined or named as such. Since the word "baseline" is seldom used in conjunction with inventories and assessments, they often are not identified as baselines. They do act as baselines, however, since they provide the data and information for evaluating the status of the environment. Another form of the "indirect" baseline is the statement of policy or opinion that forms the basis for operations and activities. For example, company environmental policies are a reflection of the corporate environmental philosophy and set the baseline for all company environmental practices. The regulatory agency mandate stated in environmental legislation forms the baseline for all agency actions. In both cases, if the record of activities matches the stated philosophy, then all operations and activities flow from one baseline. The "double baseline" or "double standard" exists when practice does not match the stated philosophy.

ESTABLISHING BASELINES

Baselines are established in a variety of ways, using one or a combination of the following:

- Empirical data and information;
- Scientific and technical data and information;
- Policy-making processes; and
- Public opinion.

Some baselines evolve as the result of the empirical accumulation of data and information. This may occur over a long period of time, as with data and information that are collected during routine operations. Often, however, in the course of the evaluation and debriefing that follows an emergency situation, valuable data and information are acquired and rapidly become the baseline against which actions in subsequent emergencies are measured. In either case, data and information are acquired in the course of performing operations and not as a part of a specific study or monitoring program carefully designed for the express purpose of collecting them. These baselines are inherently exploratory in nature and tend to be more dynamic than technically derived baselines. Adjustments of empirically derived baselines are often required and can eventually lead to the creation of completely new types of baselines. This is often the case in today's climate of rapid technical development driven by the increase in environmental regulations.

The preferred type of baseline is one that has been established using data and information gathered by scientific and technical research and conducted according to set protocols with clearly stated objectives. This type of baseline is usually numerical in nature and offers a clearly defined point of reference. Baselines founded on sound science and technology are the most difficult to establish, since they require rigorous research that may extend over long periods of time and incur substantial costs. Unless mandated by law, they may not exist, since few are willing to invest the time and money that is needed for more pressing matters. The data and information may have been collected, but not assembled or analyzed for the purpose of establishing a baseline. This is often the case with ambient air quality monitoring data. Also, because of costs involved in data collection, baselines may not have been updated and may be based on old data that are no longer valid or appropriate. In either case, if a baseline is required, work may be suspended until one is developed.

While technically derived baselines evolve from particular activity, the policy-derived baseline is a direct outcome of the policy-making process. Most policies are based on certain standards or demands of a constituency and have specific operational baselines. The baselines are established in a manner consistent with the policy-making process and are a direct result of that process. An example is the baseline set in environmental legislation to support the government policy of protecting public health and the environment. Legislative baselines are those that are specifically mentioned in the laws and regulations and have been created by legislative mandate. They are extensions of the laws that have been promulgated to address specific environmental problems and may be a combination of all of the above plus public opinion. Legislative mandates are concrete expressions of public will and, as such, are based on public perception and need.

Finally, baselines established because of public opinion should be considered. Because of vocal public concern, regulatory agencies and the private sector may find themselves in the position of revising standards or measurements to meet public demands. While these concerns may be based solely on perception, the reality is that time and money may have to be spent to establish baselines that are acceptable to affected communities. Increasingly, companies are developing more stringent standards for operations, not only as a matter of compliance and economic competitiveness, but also as good public relations. Many companies now find the use of operational baselines that are "environmentally sensitive" to be an important part of their community relations campaigns. As elected officials respond to public opinion and pressures to pass more stringent environmental legislation, this is accompanied by the development of new and more stringent baselines. These new baselines may or may not be needed, but they are brought about by public opinion.

Decisions regarding the need for baselines are usually made at the legislative or corporate level. After these requirements to establish baselines are set forth, the development is done by a variety of entities, including regulatory agencies, the scientific community, the business/industrial regulated community, and professional societies or associations.

The data and information that will compose the baseline may have already been collected in environmental monitoring programs, research laboratories, or testing facilities. If so, the baseline can be established in a relatively short time frame. It should be noted, however, that this may take a little longer if the data and information are not readily available or if its existence is not a matter of common knowledge. This is often the case where massive volumes of data have been collected over long periods by regulatory agencies and then simply archived.

Often, the data and information do not exist for the simple reason that they have never been collected or generated. If this is the case, a period of years may be required to establish a baseline. Also, a specification of the baseline may be that it covers a set period of time, in which case its development will be slow. Inventories of flora and/or fauna in an area are often required before a baseline can be established. Again, this will involve a longer period of time, particularly if seasonal population variations are expected and the inventory must be conducted over several seasons.

Some baselines are difficult to establish in terms of time and money. One such example is that of the community health effects baseline. Usually, there is a scarcity of data and information or conflicts among existing reports. Often, baselines are not considered until the population in an area has been exposed for a prolonged period of time, thus eliminating any possibility of determining what "health" baselines existed prior to the exposure. Little work has been done in this area, since most researchers prefer to devote their efforts to acquiring funding for short-term research projects that yield quick results for journal publications.

BASELINES IN THE 1990 CAA

The provisions of the 1990 CAA Amendments will significantly change both the numbers and types of baselines that will be used in environmental auditing. Environmental audits will need to be adjusted to accommodate many of the new provisions of the Amendments. Future audits will be conducted with reference not only to compliance within the facility, but to how the facility impacts the surrounding area. The 1990 CAA shifts the emphasis of compliance auditing from permitted emissions to pollution prevention. The concept of pollution prevention is based on using established norms (i.e., baselines) to comparing past, present, and future conditions.

Under the 1990 CAA, increased emphasis is placed on the limitation of emissions in relation to the State Implementation Plan (SIP), Prevention of Significant Deterioration (PSD), and attainment of the NAAQS. The provisions set forth in Title IV, Acid Deposition Control, and Title VI, Stratospheric Ozone Protection, are based on the reduction of SO_2, NO_x, and CFCs emissions, comparing current emissions against previous (baseline) units of production in specified (baseline) years. This will have a major impact on conducting audits by shifting the emphasis from permitting and compliance to pollution prevention. It also calls for auditing to be done from the perspective of preventing or eliminating those factors that constitute a potential risk, and discourages the acceptance of controlled risks.

To accomplish this change, adherence to established baselines will be the standard by which all environmental compliance audits will be conducted in the future. The provisions of the 1990 CAA have greatly expanded the requirements for establishing and using baselines, as shown in Table 8-1.

The 1990 CAA contains 143 parts or subparts in 48 sections, where requirements are set forth for establishing long- and/or short-term baselines by regulatory agencies or regulated entities. A detailed listing of baselines found in each section is provided in Appendix 8-A at the end of this chapter. As noted in Table 8-1, there are no listings for Title IX, Clean Air Research; Title X, Disadvantaged Business Concerns; and Title XI, Clean Air Employment Transition Assistance. The material found in these titles does not contain baseline requirements and was not included.

TABLE 8-1 Baseline Provisions in the 1990 CAA

Title	Number of Sections	Number of Parts and/or Subparts
I	12	29
II	9	14
III	2	16
IV	8	42
V	2	3
VI	10	32
VII	2	4
VIII	3	3

The 1990 CAA has increased requirements for establishing environmental baselines by regulatory agencies at both the federal and state levels. It also requires regulated entities to use these baselines when designing their emission reduction and control programs. Prior to beginning any air quality audit, the auditor should contact the appropriate regulatory agencies to determine what baseline data and information are available. Prudent auditors will establish data bases for the access of existing baseline data and information as early as possible, to prepare for auditing under the 1990 CAA. Since baselines requirements will vary depending on the project and client, the auditor may also wish to prepare separate listings of the baselines required for each situation. The following summary review of the baseline requirements in each title of the 1990 CAA will quickly orient the auditor to the use of baselines in conducting future compliance audits.

SUMMARY OF BASELINE REQUIREMENTS

Title I

Title I contains the "provisions for attainment and maintenance of National Ambient Air Quality Standards" and requires the establishment and use of three basic types of baselines to accomplish these objectives. As shown in Appendix 8-A, these consist of standards, inventories, and limitations. The standards and inventories are numerical and clearly stated, while the limitations are both numerical and descriptive.

The NAAQS form the baseline against which all compliance activities (with the exception of acid rain and stratospheric ozone) are measured. Auditing of the facility will be based on how emissions from the facility are controlled in order to meet or attain these standards. Permit emission limits, traffic control measures, offset reductions, and the control of VOC commercial and consumer products are all set within the context of attaining standards.

To determine what sources contribute to the ambient air levels of these substances, a baseline consisting of emission and traffic inventories is required. Emission inventories from all sources plus emission cost control measures are required for NAAQS (TSP, PM-10, CO, SO_2, NO_x, Pb, and ozone). There are also special provisions for emission inventories for VOCs, CO, PM-10, and ozone. Emission inventories from all sources include not only major and minor emission sources at facilities, but also emissions from all anthropogenic sources, including domestic sources. This implies that audits of emissions will have to be conducted by the regulatory agencies as well as facilities. Traffic inventories are also required to collect data and information on travel routes, mileage, and passenger occupancy to control

TSP, PM-10, CO, and NO_x levels to meet attainment of NAAQS, SIP, and PSD provisions. At this point, it is not clear whether inventories will be classified as part of the compliance audit, the assessment audit, or both. Hopefully, this will be clarified in the ensuing rules and regulations.

These inventories also form the baselines for the required emission reductions from major sources as well as reduction ratios and offset ratios. They are used as the basis for designing SIPs and PSD programs. The emission inventories required under Title I are used to set many of the baselines found in other titles of the 1990 CAA.

Facilities must collect, calculate, and report the quantity and quality of their emissions to the Environmental Protection Agency (EPA) and state agencies. In turn, the agencies are required to issue a report of these emissions. These inventories are long term in nature, both in terms of collection periods and time periods for which they are valid. Although their preparation may place a strain on existing resources, the information obtained in the inventory can be used to identify areas where cost savings can be implemented, such as improvement of process operations, product substitution, and hazardous waste reductions.

Data for the ambient air baseline can be obtained from the EPA and from state agencies. These data generally will be limited to the six substances that form the NAAQS; however, occasionally other data from special studies can be found. Also, facilities may have collected ambient air quality data as a part of SIP, PSD, or permit monitoring requirements.

While determining the time period over which the data were collected, the auditor should also be aware of the changes in analytical methods and quality control procedures that may have occurred during this period. The reported ambient air quality values may reflect changes in both of these. For example, when the ozone analytical method was changed, a correction factor had to be applied to standardize the data. Failure by the states to do this may be reflected in the sudden changes in ozone levels observed in the data base. Also, the auditor should check the data base carefully for "missing data." "Missing data" can be either missing individual values, such as a daily record, or an absence of data for a specific substance. An example is ambient air monitoring for VOCs. These data are limited to nonexistent because they were not NAAQS and therefore not high-priority items on the monitoring list. The auditor may find some sources for missing data in the private sector.

The auditor should carefully review past emission inventories, not only in terms of quantities and qualities of emissions, but also to determine how the data were collected and what calculations were used to arrive at the final figures. Particular attention should be given to the requirements for using calculated versus measured emissions. Calculations of emissions based on values obtained from such sources as AP-42 may yield higher values than those obtained by actual measurements. Monitoring methods should also be audited, since they can also be a source of variation in the data.

Limits set forth in Title I also form the baseline for attainment in terms of the baseline years designated, percentage reductions of VOCs from baseline values over a set time period, the reduction numbers for PM-10 reductions, and attainment levels and areas. Although these are primarily limits for ambient air quality levels, auditors should pay particular attention to these limits when conducting the compliance audits, since facility emissions will be increasingly evaluated in terms of contributions to ambient air quality.

Title II

Title II regulates both the emissions from mobile sources and the products of manufacture that can potentially contribute to a decrease in ambient air quality. It is significant in that it calls not only for the use of vehicle emission inventories, but also for the use of manufacturers sales volumes to set baselines.

The baseline in the form of a limit of ten or more vehicles is set in the definition of a covered fleet. Additional limits are set for vehicle gross vehicular weight rating (gvwr) and populations in non-attainment areas. Baseline years for NAAQS are also specified. Baselines are also set on NMHC, CO, PM-10, and NO_x emission inventories for certain vehicles with emission reductions based on manufacturers' sales volumes. Baseline vehicles and baseline gasolines are specified. There is an underlying assumption that these requirements will affect only vehicle manufacturing plants and those state agencies charged with the development of SIPs. It has, however, far-reaching implications and may require additional facility inventories of vehicle fleets.

Title III

Although Title III, Air Toxics, has only two sections (Sections 301 and 305) that set forth requirements for baselines, they contain 16 parts and subparts that will have far-reaching consequences in both auditing and resource utilization. These baselines consist of a list of substances, inventories of sources, emissions, and accidental releases, baseline years, and data and information on health effects and atmospheric deposition. Although much of the data and information may have been collected in some form by agencies and facilities, as yet little has been done to assemble it in the form needed to establish the required baselines. The establishment of baselines required under this Title will be expensive and time-consuming, and prudent companies will quickly initiate comprehensive, aggressive hazardous air pollutants (HAPs) auditing programs to collect the required data and information.

The first baseline for toxic substances is clearly set in Section 301 (b) (1) in the listing of 189 HAPs. Although the matching of chemicals contained on site with this list appears to be rather straightforward, the auditor should give careful attention to the contents of process mixtures, by-products, and waste.

Emission inventories must be completed and all area sources identified within 5 years of the passage of the 1990 CAA, since the EPA is required to list the area sources that pose the greatest threat to human health and the environment. This listing will be based on the area source emissions (actual or estimated aggregates) that comprise 90% of the 30 HAPs that present the greatest threat to human health and the environment. The baseline for emissions from new sources will be the average emission limitation achieved by the best performing 12% of existing sources. Early reductions will also be based on the emission inventories of HAPs and HAP particulates. Again, although the regulatory agencies will bear the responsibility for assembling data and information, preparing the reports to Congress, and sometimes even conducting studies, the collection part will fall mainly on the facilities. The initial facility audit of HAPs emissions and sources can be done relatively quickly, since much of the data and information have already been collected under the SARA, Title III reporting requirements.

Baselines for health and environmental effects will undoubtedly be the most difficult and time-consuming. Although the short-term health effects of many substances is well documented, data and information concerning long-term health effects may be difficult to obtain. The auditor should start by collecting all data and information available at the facility and in other company data bases. The next step would be to consult other sources, such as local and state health departments, the Centers for Disease Control (CDC), and the Agency for Toxic Disease Registry (ATSDR). Computerized literature data bases, such as TOMES in MEDLINE and literature abstracts, are also excellent sources of information. These sources, however, may not provide data or information relevant to the area of interest. Data and information regarding human health effects may exist in hospital and physician records, but these sources will be almost impossible to use because of record confidentiality. In addition to these, other sources of data and information regarding environmental effects on flora and fauna include the state and federal agricultural and fish and wildlife agencies. Since most companies will not be able to conduct health effects studies because of the expense and long time periods involved, the responsibility for these will, probably be borne by federal agencies. Finally, health and environmental effects data will probably be assembled in an assessment type audit rather than a compliance audit.

The identification and assessment of the extent of HAPs deposition in the Great Lakes, the Chesapeake Bay, Lake Champlain, and coastal waters will be a major project that will not be completed rapidly. This project, however, will obtain information and data necessary to establish a baseline for HAPs deposition in each of these bodies of water. At this point, it is unclear how much work will be assigned to the private sector and how much to federal agencies.

Section 301(n) (1) requires that studies of mercury (Hg) emissions from electric steam generating units (ESGU), municipal incinerators, and other area sources be conducted. These studies must include information about emission rates, control technologies, the cost of controls, and health and environmental effects. Again, these studies will be expensive and lengthy. The implication is that federal agencies will take the lead in conducting the studies; however, the private sector may be required to participate. Auditors should be particularly alert for relevant data and information that may have been already collected at a facility.

Section 301(n) (3) contains the provision for studies to be conducted to characterize HAP emissions from publicly owned treatment works (POTW). This will include identifying industrial, commercial, and residential discharges that contribute to these emissions. Again, much of this information and data will be available at the facility; however, the auditor should be aware that new auditing procedures may be required to meet additional requirements imposed by this section.

Perhaps the most extensive project will be the one called for in Section 301(n) (4). This requires that all area sources of oil and gas wells and pipeline facilities be inventoried in metropolitan areas with populations of over 1 million. Although this will probably be done by agencies, auditors should be aware that facilities and energy companies may be required to furnish complete oil and gas well and pipeline inventories.

Inventories of emissions from solid waste combustion required under Section 305(a) (2) will be used to form the baselines for emission standards for these units. Average emission limitations will be based on the best performing 12% of the units in the specified categories.

Although Title III does not contain the most sections, parts, and subparts, it does contain substantial requirements to establish baselines. Some data and information may exist, but much is scattered and will be difficult to assemble. The required studies will be costly, lengthy, and difficult. The extent of private sector involvement is unclear, particularly in view of diminishing agency resources and the notable absence of references in the law for funding of the studies.

Title IV

Title IV, Acid Deposition Control, contains the largest number of provisions for establishing baselines. It consists of 8 sections and 42 parts or subparts.

The majority of these regulate utility industries; however, other industries may be affected by certain provisions of the acid deposition sections (404 and 405).

Baselines in this section are set using emission inventories (SO_2, NO_x), fuel consumption inventories, the identification of environmentally sensitive areas, allowance allocations, defined production units, and limits set forth in NAAQS and SIPs. Baseline years are specified in all sections, with the exceptions of Sections 407 and 415.

Section 402 defines "baseline" as the annual quantity of fossil fuel consumed by an affected unit, measured in millions of British thermal units (mmBtus). This is a major change and sets a precedent for establishing baselines on quantities of renewable resource consumption. A "utility unit" is defined as a unit that, during 1985, served as a generator in any state that produced electricity for sale. It further designates which years are to be used when calculating baselines for utility and nonutility units.

Sections 404 and 405 collectively contain 19 parts and subparts in which baselines are defined. Phase I SO_2 requirements are set forth in Section 404. Baseline years are designated for permit applications and for establishing actual and allowable emission rates and allowance allocations. Section 405 contains the Phase II SO_2 requirements. The baselines contained in this section apply primarily to allowances and are set on the basis of the type of unit, its electrical generating capacity, the type of fuel used, fuel consumption, emission rates, and year when the unit began operations. Baseline years are set in each section.

The amendments to Sections 404, 405, and 408 pertain to acid deposition and are found at the end of Title IV. A baseline consisting of the identification of sensitive and critically sensitive aquatic and terrestrial resources in Canada and the United States that may be affected by the deposition of acid compounds is required under Section 404. This is followed in Section 405 by the requirement of the EPA to establish a listing of all acid lakes. Finally, Part 408 contains provisions for monitoring an acid rain program in Canada. It calls for the analysis of current SO_2 and NO_x emission levels in each of the Canadian provinces, the costs and methods used to achieve reductions, and environmental impacts. The results of this analysis are far reaching, since they will form the baselines for future acid rain programs. Currently, these baselines fall in the assessment audit category rather than the facility compliance audit category.

The NO_x Reduction Program baselines in Section 407 for affected boilers are set on mmBtus and NO_x emission rates. Units are defined the same as those subject to SO_2 reduction requirements.

The next three sections apply to facility compliance. Permit Compliance Plans (Section 408) must be in compliance with both NAAQS and calculated unit allowances based on 1985 emission rates. Repowered sources annual SO_2 allowances are based on 1995 actual emission rates in Section 409. In Section 410, part 410 allowance baselines established by the EPA are to be based on fuel consumption and operating data for the designated baseline years. An amendment to Section 409 requires the U.S. Department of Energy (DOE) and Department of Commerce (DOC) to inventory and analyze clean coal technology export programs and report the findings to Congress within 1 year of the enactment of the 1990 CAA.

Finally, in Section 410, baselines are set over a 3-year period for additional sources, based on fuel consumption and unit operating data.

Title V

Title V contains only two baseline requirements. Since these are implied, the auditor may find it difficult to identify them as such. In Part 502(b)(3)(B), the permitted amount of regulated pollutant forms the baseline for the fees imposed for excess emissions. The second baseline (Section 502,c,5,B) limits permits to 5 years.

Title VI

Title VI contains ten sections and 32 parts and subparts and has the second largest baseline requirements in the entire act. It will affect not only regulated facilities, but also the public in the provisions for stratospheric ozone protection and the regulation of chlorofluorocarbons (CFCs). It will also require auditors to make new additions to their existing compliance audit procedures.

Section 601 defines the baseline years for Class I, II, III, and IV substances. It also defines "consumption" as equaling the amount of these substances produced in the United States, plus the amount imported, minus the amount exported to parties in the Montreal Protocol. It sets the chlorofluorocarbon-11 (CFC11) as the substance against which ozone depletion will be measured and as the basis for measurements.

Section 602 contains the listing of Class I with Groups I–V and Class II substances. This list forms the baseline of substances regulated under this title. It also sets a compliance deadline for newly listed substances as 7 years.

The monitoring and reporting requirements of Title VI are contained in Section 603 and are based on production, use, and consumption of Class I and II substances. It also calls for the National Oceanic and Atmospheric Administration (NOAA) and the National Aeronautics and Space Administration (NASA) to report to Congress on the levels of stratospheric ozone depletion every 3 years. This will include information on bromine (Br) and chlorine (Cl) levels, peak Cl loading, and stratospheric ozone depletion, as well as a projected date for reducing atmospheric levels of Cl to 2 ppb. Finally, the base case sets 2000 as the target year for the global phaseout of all hydrocarbon emissions. While this section places major monitoring responsibilities on federal agencies, it also assigns all facilities producing and using Class I and II substances additional reporting responsibilities. Again, auditors will want to make adjustments in their auditing protocols to include procedures that will evaluate compliance in this area.

The phaseout of the production and consumption of Class I substances is found in Section 604 and of Class II substances in Section 605. Section 604 contains a table of baseline years for limiting the percentage of maximum allowable production, while Section 605 specifies dates and quantities for producing Class II substances beyond the baseline year. Section 606 calls for an accelerated phaseout schedule to be established, based on assessments of the harmful effects of Class I and II substances on the ozone layer. This will require that a baseline be established against which effects can be measured to determine what is or is not harmful.

Sections 608, 609, 610, and 612 will directly affect the private as well as the public sectors. Under the national recycling and emission reduction program set forth in Section 608, no maintenance, servicing repair, venting, or disposal of Class I or II substances for appliance or industrial process refrigeration substances will be allowed after January 1, 1992. Audits performed after the baseline date will need to include a check of compliance with this provision. Sections 609 and 610 will affect the domestic as well as commercial sectors, since it sets baseline years after which no servicing or repairs may be done to automobile air conditioning units unless they are performed by certified individuals according to specifications. A baseline date for submitting certification is also set. Both the specifications and the standards for certification set the baselines for conducting such work. Facilities with fleets of vehicles and businesses that are engaged in automobile air conditioning repair and servicing should be certain that a complete compliance audit is conducted for the requirements contained in these sections.

Section 610 sets the baseline for restrictions on nonessential products containing CFCs by defining and identifying these products. The auditor should cross-check all substitutions with respect to potential regulation under other environmental legislation.

Section 612 requires that the producers of any substitute for Class I substances submit any published or unpublished health and safety study data. These data will undoubtedly be used to develop future baselines for health effects of these substances.

Information and data collected in the inventories and studies required in Section 603 amendments will be used to form the baselines for the development of future methane regulations. These requirements are extensive in that they call for inventories to be made of natural gas, coal extraction, and solid waste domestic methane sources, including the storage, transportation, use, and distribution for each. Also included are inventories of methane production from other domestic sources, including the agricultural production of rice and livestock, the biomass burning of agricultural wastes, wood, grasslands, and forests, and other human activities. These requirements will certainly increase auditing requirements in the public sector for local, state, and federal agencies. Given the diminished allocation of resources to these agencies, it is reasonable to expect that much of the burden of these audits will be passed on to the private sector; therefore, auditors should be prepared.

Title VII

Title VII, Provisions Relating to Enforcement, contains only two sections (701 and 702) in which baselines are set forth. First it sets the use of the 1986 IRS Code as the baseline for civil penalties. It also lists the basis for penalty assessments as size of the business, economic impacts of the penalty on the business, compliance history including good faith efforts, duration of the violation, payment of previous penalties for the same violation, economic benefits of noncompliance, and the seriousness of the violation. Since compliance baselines and requirements are set forth in the previous titles, much of the data and information will have been collected in previous audits or in other sections of the current audit. Section 702 requires that owners and operators of emission sources and manufacturers of emission control and process equipment establish and maintain complete records and reports of monitoring and production. Auditing ensures that these requirements are met and forms the basis for the compliance certification.

Title VIII

The final set of baselines are contained in Title VIII, Miscellaneous Provisions, which has three sections: 801—Outer Continental Shelf (OSC) Air Pollution; 816—Visibility; and 817—the Role of Secondary Standards. These provisions require agencies to conduct inventories and assessments that will provide data and information that can be used in the future to set baselines.

The baseline in Section 801 consists of a definition of offshore areas and is rather straightforward in its specifications. Much of the data and information required by Section 816 for agency assessment of current sources of visibility impairing pollution and clean air corridors can be found in previous emission inventories. These are important in that they provide for the collection of information and data that potentially will be used to set future visibility baselines. Facilities should be prepared to respond to agency requests for data and information that have been collected or evaluated in previous facility audits.

Finally, Section 817 calls for a determination of ambient air concentrations that would be adequate to protect welfare and the environment from the effects of pollutants listed in Section 108. These provisions will set the welfare and environmental effects baseline for secondary standards. Any future auditing should be expanded to include data and information related to these requirements.

ADJUSTMENTS IN AUDITING UNDER THE 1990 CAA

The range of requirements to establish baselines and the baseline definitions found in the 1990 CAA clearly point to the increase in the importance of baselines in the auditing process. The 1990 CAA not only makes extensive use of baselines established in previous legislation, but it requires the development of many new baselines. Facility owners and operators may find that not all of these will directly affect their auditing procedures; however, they should be aware that regulatory agencies may request audit data and information to assist them in establishing certain baselines.

The 1990 CAA also sets several precedents in baseline development. First, it uses process performance rather than ambient air quality standards as the baseline for compliance. Second, it establishes baselines on resource consumption units rather than on environmental quality. Third, it introduces the concept of developing baselines for entire regions. This requires that environmental audits be conducted over large regions to obtain data and information to use in setting baselines for the region. There is also the implication that baselines may become more regionally sensitive—that is, different baselines for different regions and circumstances. The 1990 CAA also sets baselines almost on a case-by-case basis until the wording of the law begins to resemble that found in the final rules and regulations.

Finally, the 1990 CAA greatly expands environmental auditing via baseline requirements and the emphasis on compliance auditing. Appendix 8-A gives ample proof of the wide range of baselines used in the 1990 CAA and new ones developed to meet its requirements. The auditor can expect to see many of the types of baselines estblished by this act in future environmental legislation. This will continue to expand the audit, as baselines are developed to address new areas of concern. The increase in baselines that are developed over long periods of time only reinforces the case for increased importance of baselines in environmental auditing.

In an effort to prepare to meet the requirements of the 1990 CAA, environmental auditors should identify those baselines that are required for specific audits. Multiple readings of the 1990 CAA will not only familiarize the auditor with its contents, but will begin to give the auditor an idea of the extent of baseline requirements. Summaries of the 1990 CAA, such as that given in Appendix 8-A, will be of assistance to the auditor when determining what baselines will be required and the adequacy of existing baselines. The auditor should list the baselines currently used for each type of facility audit and determine their adequacy to meet the requirements of the 1990 CAA. Working closely with clients, regulatory agencies, academia, and professional associations, environmental auditors can begin to develop and establish baseline registries. Ready access to available baseline data and information will be the key to successful future audits.

References

Clean Air Act Amendments of 1990, PL 101-549, November 15, 1990.

Appendix 8-A

Baseline Requirements

Title	Section	Baseline
I National Ambient Air Quality Standards	172 NAAQS	(c)(3) qualify and quantify emissions from construction and existing/new/modified sources
	173 Permits	(c) baselines of NAAQS for offsets (e) emission inventory of rocket engines and motors (e)(4) emission fees to be based on average cost × 1.5 of stationary source control measures over last 3 years
	179 Non-attainment and Sanctions	(b)(2) 2:1 emission reduction requirements for modified sources or emission units (c)(2) identification of non-attainment areas and existing NAAQS levels (qualitative, quantitative, temporal, spatial)
	181 Attainment Dates	(a) 1988 and 1989 baseline years (a)(4) area reassignment if it has ± 5% of level on which classification was based (a)(5)(B) number of exceedances per year
	182 SIP	(a)(1) emission inventory from all sources (a)(4) ratios for offset requirements—1:1 (total VOC emission reductions to total increased VOC emissions) (b)(1)(B) baseline for VOC and NO_x emissions from all anthropogenic sources (b)(1)(C) 15% reduction in VOC from 1990 levels within 6 years (b)(1)(D) RACT requirements for all major, stationary VOC sources in the area (c)(2)(B) reasonable further progress demonstration of at least 3% reduction of VOC baseline emissions/year and <3% if technically feasible (c)(3)(C)(5) transportation control includes inventory of current aggregate vehicle mileage; aggregate vehicle emissions; traffic congestion levels (d)(1) inventory of vehicle trips; miles traveled by employees; travel patterns; average passenger occupancy (d)(2) general offset reduction requirements of 1.2:1 (total VOC emissions:total VOC increased emissions) (d) and (e) VOC reduction ratios (1.2:1, 1.3:1, 1.5:1)
	183 Ozone	(e)(3) 80% reduction in reactivity adjusted on basis of VOC commercial and consumer products

Title	Section	Baseline
	185 Severe and Extreme Areas Ozone Non-attainment Areas	(b)(2) guidelines based on lower of average of actual or allowable values determined over 1 calendar year
	187 CO SIP	(a)(1) inventory of all emission sources (a)(5) periodic inventory every 3 years (b)(2) vehicle miles traveled—SIP revisions
	188 Classification and Attainment Dates for PM-10	(e) numbers and types of PM-10 sources or other PM-10 emitting activities in the area
	189 PM-10 SIP	(b)(3) 70 tpy major sources in serious area (d) ≤5% annual reduction of area emissions based on last emissions inventory
	191 SIP Submission Deadlines	Within 18 months of enactment of 1990 CAA
	192 AttainmentDates	(a) ≤5 years of non-attainment designation
II Mobile Sources	203 Emission Standards for Conventional Motor Vehicles	(a) NMHC, CO, PM-10, and NO_x emissions for LDT models after 1993, up to 6,000 lbs gvwr based on manufacturers' sales volumes
	204 CO Emissions at Cold Temperatures	(1) emission reductions based on manufacturers' sales volumes
	216 Fuel Volatility	(a)(1) based on Reid Vapor Pressure of ≤9.0 psi in non-attainment areas
	217 Diesel Fuel Sulfur Content	(1) ≤0.05% sulfur by weight or centane index ≤40

Title	Section	Baseline
	219 Reformulated Gasoline and Oxygenated Gasoline	(k)(4) baseline gasoline—gasoline sold or introduced into commerce in 1990 (k)(9)(A) baseline vehicle—1990 model
	227 Urban Bus Standards	(b)(1) 50% reduction in PM-10 levels based on 1990 emissions or 70%; vehicle model year—1994; reductions based on durability, cost, safety, lead time, and other relevant factors (c)(2) low-polluting fuel requirement in areas with a 1980 population of 750,000
	241 Definitions	(5) covered fleet—≥ 10 vehicles owned or operated by a single person
	242 Clean Fuel Vehicles	(b) clean fuel vehicles of 8,500 gvwr (c)(2) LDTs and trucks of 6,000 lbs gvwr
	246 Centrally Fueled Fleets	(a) baselines for fleet programs required in CO and ozone non-attainment areas (a)(2)(A) in 1987, 1988, and 1989 ozone non-attainment areas with 1980 population of ≥ 250,000 (a)(2)(B) in 1988 and 1989 CO non-attainment areas with ≥ 16 ppm and a 1980 population of ≥ 250,000
III Air Toxics	301 Hazardous Air Pollutants	(b)(1) list of 189 hazardous air pollutants (HAPs) (c)(3) 5 years after passage of 1990 CAA, the EPA shall list the area sources that pose the greatest threat to human health and the environment, based on actual or estimated aggregate emissions representing 90% of area source emissions of the 30 HAPs that present the greatest threat to human health and the environment (d)(3) new sources—based on average emission limitation achieved by best performing 12% of existing sources (f)(1)(C) baseline on health effects derived from existing data, epidemiologic studies, and other health studies (i)(5)(A) early reduction demonstration that a source has reduced 90% of HAPs (95% of HAPs particulates) for 6 years of compliance data for an applicable standard (i)(5)(C) baseline year ≥ 1987 for verifiable and actual emission reductions or a baseline year of

Title	Section	Baseline
		1985 or 1986, provided that data is received by the EPA prior to enactment of 1990 CAA (k)(1) reductions of 75% of incidence of cancer attributable to emissions from area sources (k)(3) identification of area emissions of 30 HAPs that present the greatest public health risk/threat; identification of sources producing 90% of the aggregate emissions of each of the 30 HAPs (m)(1) identification and assessment of extent of HAP deposition in Great Lakes in Chesapeake Bay, Lake Champlain, and coastal waters (n)(1)study of Hg emissions from EUSGU, municipal incinerators, and other area sources (rate and mass of emissions; health and environmental effects of emissions; technologies available to control emissions; cost to control emissions (n)(1) NIEHS to study and determine threshold levels of Hg exposure below which adverse health effects are not expected to occur (n)(3) studies to characterize HAP emissions from POTW and identify industrial, commercial, and residential discharges that contribute to these emissions (n)(4) area sources of oil and gas wells and pipeline facilities in consolidated metropolitan areas of > 1 million people (r)(6)(S) annual report to Congress by the Chemical Safety Board of Information of accidental releases reported during previous year (r)(7) a release history of regulated substances for the past 5 years, including size, concentration, duration, and worst case accidental releases
	305 Solid Waste Combustion	(a)(2) emission standards $\leq$ average emissions limitation achieved by the best performing 12% of units in the category (excluding units that first met LAER 18 months before the standard was proposed or 30 months before standards were promulgated)
IV Acid Deposition Control	401 Findings and Purposes	(b) reduce SO_2 emissions by 10 million tons of 1980 emission levels and reduce NO_x emissions by 2 million tons of 1980 emission levels
	402 Definitions	(4) "baseline"—annual quantity of fossil fuel consumed by an affected unit, measured in millions of British thermal units (mmBtus) (4)(A) baseline for units in operation prior to 1985—annual average quantity of mmBtu

Title	Section	Baseline
		consumed in fuel during 1985, 1986, and 1987, as recorded by DOE Form 767
		(4)(A) baseline for utility unit for which no form was filed—level specified in the 1985 National Acid Precipitation Assessment Program (NAPAP) Emission Inventory, Version 2, National Utility Reference File (NURF), or in a corrected data base established by USEPA
		(4)(A) baseline for any other nonutility unit not included in NAPAP emissions inventory, Version 2, or a corrected data base—annual average quantity in mmBtu consumed in fuel by that unit, as promulgated in regulations (not later than 18 months after passage of 1990 CAA)
		(17)(A)(ii) "utility unit"—a unit that during 1985 served a generator in any state that produced electricity for sale
	404 Phase I SO_2 Requirements	(a) 1995 will be baseline year for reductions of total tonages from all units
		(d)(2)(c) baseline of application for permit extension calculated on actual 1985 emissions rate, allowable 1985 emissions rate, and projected utilization for calendar years 1995–1999
		(e)(1)(A) SO_2 emission reductions based on 1980–1985 emissions, also 1985–1987 emissions
		(e)(2) allowances for early reductions based on 1985 actual SO_2 emission rates
		(h)(1)(2) allowance allocations for units in certain utility systems, with optional baseline based on 1989 average annual emission rates
	405 Phase II SO_2 Requirements	(a) baseline—1985 emission rates for utility units
		(b) baseline—units $\geq$ 75 Mwe and 1.20 lbs/mmBtu
		(b)(2) allowance allocations based on actual 1985 emission rates
		(b)(3) baseline for SO_2 emission tonnages—annual average fuel consumption during 1985, 1986, and 1987
		(b)(4) baseline for allowances—1988 electrical generating capacity of 30 million Kw or conversion to coal between 1/1/80 and 12/31/85
		(c)(1–5) baseline for allowances and SO_2 emissions for coal- or oil-fired units >75 MWe and >1.2 lbs/mmBtu—1985 emission rates and aggregate nameplate fossil fuel steam-electric capicity as of 12/31/89
		(d)(1–5) baseline for allowances and SO_2 emissions

Title	Section	Baseline
		for coal-fired units >1.2 lbs/mmBtu—formula derived using lesser of actual or allowable 1985 emission rates (e) baseline for allowances and SO_2 emissions for oil- and gas-fired units >.60 lbs/mmBtu and <1.2 lbs/mmBtu—formula derived using lesser of actual or allowable 1985 emission rates (f)(1–2) baseline for allowances and SO_2 emissions for oil- and gas-fired units ≤0.60 lbs/mmBtu—formula derived using 1985 emission rates and 1980–1989 annual average fuel consumption (g)(1) baseline for allowances and SO_2 emissions for units that commenced operations between 1987 and 12/31/95—formula derived using 1985 emission rates (g)(2–5) baseline for allowances beginning 1/1/2000 for Units that commenced operations between 1986 and 12/31/95—formula derived based on date construction began, date commercial operations began, and unit's annual fuel consumption (h)(1–3) baseline for allowances beginning 1/1/2000 for oil- and gas-fired units <10% oil consumed—formula derived based on 1985 emissions rate and 1980–1989 annual fuel consumption (i)(1–2) baseline for allowances beginning 1/1/2000 for oil- and gas-fired units <10% oil consumed—formula derived based on 1980–1988 population growth, 1988 electrical generating capacity, and lesser of actual or allowable 1980 emissions rate (j) baseline for allowances beginning 1/1/2000 for certain municipally owned power plants—formula derived based on lesser of actual or allowable 1985 emission rates and annual fuel consumption
	406 Allowances for States with Emission Rates at or Below 0.80 lbs/mmBtu	(a) baseline year for allowances—1985
	407 NO_x Reduction Program	(a) units subject to SO_2 reduction requirements (b)(1) affected boilers: tangentially fired boilers of 0.45 lb/mmBtu; dry bottom wall-fired boilers of 0.50 lb/mmBtu (d)(2) baseline of monitoring data of equipment operations for 15 months (e) alternative contemporaneous annual emission limitations petition based on lesser of actual annual emission rate of NO_x lbs/mmBtu averaged over the units or Btu weighted average of annual emission rates for the same units if they had been operated

Title	Section	Baseline
		at the same time in compliance with applicable emission rates
	408 Permits and Compliance Plans	(b)(1) compliance with NAAQS (c)(1)(B) unit allowances calculated according to formula and based on lesser of actual or allowable 1985 emission rates
	409 Repowered Sources	(c) annual SO_2 allowances based on 1995 actual emission rates
	410 Election for Additional Sources	(b) allowance baselines are established by the EPA and based on fuel consumption and operating data for the unit of calendar years 1985, 1986, and 1987 or, if such data are not available, on alternative representative data
	415 Clean Coal Technology Regulatory Incentives	(b)(2) temporary projects must be in compliance with SIP and NAAQS and in operation for less than 5 years
	404 Acid Deposition Standards (Amendments)	(1) identification of sensitive and critically sensitive aquatic and terrestrial resources in the United States and Canada that may be affected by deposition of acidic compounds (2) description of nature and numerical value of a deposition standard or standards that would be sufficient to protect such resources
	405 National Acid Lakes Registry (Amendments)	listing of all acid lakes by the EPA
	406 Industrial SO_2 Emissions (Amendments)	(a) report of all existing inventory information by 1995 and every 5 years for the next 20 years, plus identification of units subject to Sec. 505(g), trends, and estimates of actual emission reductions (b) 5.60 million–ton cap on emissions
	408 Monitor Acid Rain Program in Canada (Amendments)	(b) analyze current SO_2 and NO_x emission levels in each of the provinces, amount of emission reductions, methods used to achieve reductions, cost, and environmental impacts
	409 Report on Clean Coal Technologies Export Programs (Amendments)	DOE and DOC will identify, inventory, and analyze clean coal technology export programs and report findings to Congress within 1 year of the enactment of 1990-CAA

Title	Section	Baseline
V Permits	502 Permit Programs	(b)(3)(B) fees of ≥ $25/ton of regulated pollutant emitted above the permitted amount (c)(5)(B) permits issued for 5 years
	507 Small Business Stationary Source Technical and Environmental Compliance Assistance Program	(c)(3)(A) "small business" defined as: employs ≤ 100 individuals; is not a major source; does not emit ≥ 50 tpy of any regulated pollutant; emits ≤ 75 tpy of all regulated pollutants
VI Stratospheric Ozone Protection	601 Definitions	(2) baseline year: (2)(A) 1986 for Class I, Group I or II (2)(B) 1989 for Class I, Group III or IV (2)(C) representative calendar year when substance added to Class I or II (6) consumption equals the amount produced in the United States, plus the amount imported, minus the amount exported to parties in the Montreal Protocol (10) chlorofluorocarbon-11 (CFC11) is the substance against which all ozone depletion will be measured on a mass/kg basis using a substance's: • atmospheric lifetime • molecular weight of Br and Cl • ability to be photolytically disassociated • other factors determined to be an accurate measurement of relative ozone depletion potential
	602 Listing of Class I and Class II Substances	(a) listing of Class I and Groups I–V substances (isomers are also regulated) (b) listing of Class II substances (isomers are also regulated) (d) compliance deadline for newly listed substances may not exceed 7 years after substance is on or is added to the list (e) ozone-depleting potential of 25 substances listed in table
	603 Monitoring and Reporting Requirements	(b) quarterly reports required for quantities and qualities produced, imported, and exported (c) reports required on quantities of Class I substances present in baseline year (d)(1) report to be issued by the EPA every 3 years on production, use, and consumption of Class I and Class II substances

Title	Section	Baseline
		(d)(2) every 3 years NOAA and NASA must report to Congress on the average tropospheric concentrations of Br, Cl, and levels of stratospheric ozone depletion and updated projections of: (A) peak Cl loading (B) rate of projected decrease in atmospheric Cl abundance after 2000 (C) projected date when atmospheric abundance of Cl will return to 2ppb (d)(2)(C) base case—controls supplemented by year 2000 targeted as the global phaseout of all hydro-carbon emissions
	604 Phaseout of Production and Consumption of Class I Substances	(a) limit of % of maximum allowable production in baseline year given in Table 2
	605 Phaseout of Production and Consumption of Class II Substances	(a) 1/1/2015 deadline for production of any Class II substance beyond the baseline year amount (d)(2)(B) during 1/1/2015–2030 no quantities of Class I substances may be produced in quantities > 110% of the baseline year production quantities; after 1/1/2030 no quantities of Class I substances may be poduced >115% of baseline year production quantities
	606 Accelerated Schedule	(a) established phaseout schedule based on the following: (a)(1) assessment of credible current scientific information regarding harmful effects of Class I and Class II substances on stratospheric ozone layer (a)(2) availability of substitutes—technological achievements, safety, and other relevant factors (a)(3) modification of Montreal Protocol for reductions
	608 National Recycling and Emission Reduction Program	(c) after 1/1/92 no maintenance, servicing, repair, venting, or disposal of Class I or Class II appliance or industrial process refrigeration substances De Minimis not subject to prohibitions
	609 Servicing of Motor Vehicle Air Conditioners	(b)(2)(A) "approved refrigerant recycling equipment"—standards must be at least as stringent as the Society of Automotive Engineers (SAE) standard J-1990 (b)(3) "proper using"—SAE Standard J-1989

Title	Section	Baseline
		(b)(4) "properly trained"—SAE Standard J-1989 certification and training under the National Institute for Automotive Service Excellence (ASE) and the Mobile Air Conditioning Society (MACS) (c) after 1/1/92 no repairs or servicing may be done unless by certified and trained persons on auto air conditioning, or 1/1/93 for places servicing <100 vehicles/year in 1990 (d)(2) 1/1/93—date for submitting certification
	610 Nonessential Products Containing Chlorofluorocarbons	(b) nonessential products defined as: (b)(1) plastic party streamers and noise horns (b)(2) cleaning fluids for noncommercial electronic and photographic equipment (b)(3) consumer products determined by the EPA to be Class I releases or nonessential (d) effective 1/1/94, other products classified as nonessential: any aerosol plastic foam product containing or manufactured with a Class II substance
	612 Safety Alternatives Policy	(e) any published or unpublished health and safety study data on a substitute for a Class I substance required from producer of substitute
	603 Methane Studies	(b)(1) domestic methane source inventory of natural gas extraction, transportation, distribution, storage, and use (b)(2) domestic methane source inventory of coal extraction, transportation, distribution, storage, and use (b)(3) domestic methane source inventory of solid waste transportation, distribution, and storage (b)(4) domestic methane source inventory of agriculture (rice and livestock production) (b)(5) domestic methane source inventory of biomass burning (agricultural wastes, wood, grasslands, and forests) (b)(6) domestic methane source inventory of methane emissions associated with human activities (d) report by the EPA to Congress not later than 2 years after passage of 1990 CAA on the estimated emissions from biogenic sources, forests, tundra, and wetlands
VII Provisions Relating to Enforcement	701 Federal Enforcement	(d)(5)(B) use of 1986 IRS code to set civil penalties (e)(1) penalty assessment shall be based on: • size of business • economic impact of penalty on business • compliance history of violator

Title	Section	Baseline
		• good faith efforts to comply • duration of violation as established by evidence • payment of previous penalties for same violation • economic benefits of noncompliance • seriousness of violation
	702 Compliance Certification	(a)(1)(A) owners/operators of emission source or manufacturers emission control or process equipment required to establish and maintain records and reports (a)(1)(E) records required on control equipment parameters, production variables, or other indirect data when direct monitoring is not possible
VIII Miscellaneous Provisions	801 OSC Air Pollution	(Sec. 328)(a)(1) offshore areas of United States defined as Pacific and Atlantic and Arctic coasts; Florida east at 87° 30' latitude; 25 miles of seaward boundary of states
	816 Visibility	(Sec. 169B)(a)(1)(B) assessment of current sources of visibility imparing pollution and clean air corridors
	817 Role of Secondary Standards	(a)(4) determination of ambient concentrations of pollutants (listed in Sec. 108) that would be adequate to protect welfare and the environment from effects

9

Air Permit Audits

Jacqueline Shields and Paul D. Rosenstock

GENERAL REQUIREMENTS FOR AIR PERMIT AUDITING

The environmental compliance auditing of air permits is perhaps one of the most important services that an auditor can perform for a company. Permitting is a complex, costly process that companies must endure if they are to legally conduct business in the United States. Most larger companies have a staff that does all the company permit work, while smaller companies usually rely on the services of a consultant to help them secure their permits. The auditor reviews and inspects records and operations to determine the status of the company with respect to compliance with the terms of the permit.

Many operations in a facility have more than one type of permit, since they produce emissions, discharges, and wastes that are regulated under different environmental laws. For example, a facility may hold a permit for the treatment, storage, and disposal (TSD) of hazardous waste under the 1976 Resource Conservation and Recovery Act (RCRA). The TSD operations may produce sufficient volumes of substances regulated under the 1977 Clean Air Act (CAA) Amendments to require that an air permit also be obtained for these operations. Concurrently, it may have another permit regulating the discharge of wastewater under the Clean Water Act (CWA). Volatile Organic Compound (VOC) emissions from the wastewater treatment process may require an air permit be obtained for this operation. Each permit is issued in conformance with the regulations set forth for the governing body of legislation and the requirements of the permitting agency. The permitting process consists of three phases: application, maintenance, and renewal.

The permit application phase occurs when a new facility is being built or when the operations are modified so significantly that the regulatory agency deems a new permit application more appropriate. Also, when new environmental legislation is enacted, it may be necessary to apply for a new permit to comply with the ensuing regulations.

The maintenance phase of the permitting begins after the permit has been issued and the facility must comply with the operational limits of the permit. The renewal phase occurs when the time limits of the existing permit have expired or when new regulations require that a renewal application be filed.

Depending on the activities and operational status of the facility, several types of permits may be required. These may include one or all of the following:

- A preconstruction permit;
- A construction permit;
- An operation permit;
- A special permit; and
- PSD and/or NSR.

Although these permits regulate different activities, all permit applications first must be submitted to the appropriate regulatory agency for review of completeness and technical accuracy. Thus, the applications for the permit may be in various stages of review by the permitting agency. Existing permits were probably issued at different times. For all types of permits, the auditing process involves certain common elements divided into two distinct phases: the administrative review and the compliance review. The checklist given in Table I shows the items in each of these two categories that should be examined in an air permit audit.

ADMINISTRATIVE AUDIT

The first phase of the permit audit process consists of an administrative review of the permit. This is essentially a review of data, information, the permits themselves, and other miscellaneous items listed in Table 9-1. Air permits should be audited for all of the items in Table 9-1 with respect to:

1. Accuracy and completeness;
2. Errors and problems; and
3. Omissions.

Although Table 9-1 is presented here in the form of a checklist, it can easily be modified to express the results of the audit in any manner desired. For

TABLE 9-1 Air Permit Audit Checklist

Administrative Audit
Inventory of existing permits
Inventory of permit applications
Status of permit application
Determination of permit renewals
Identification of the agency issuing the permit
Verification of the identity of the permit holder
Completeness of permit
Inventory of regulated substances (NAAQS, NESHAPs, VOC, HAPs)
Inventory of emissions (sources, quantity, quality)
Record keeping requirements
Reporting requirements (including those for self-reporting)
Compliance schedule
Compliance record (notices of violation, fines, etc.)
Compliance test data
Records for operation and maintenance upsets
Logs of systems, processes, and equipment operation
Records for monitoring (continuous and periodic), sampling, and testing programs
Quality assurance data and information for monitoring, sampling, and testing
Modeling requirements and results
Documentation of training, safety, and worker-right-to-know programs
Review of worker and environmental health records
Records of community concerns and complaints
Documentation of Emergency Response Plan
Documentation of pollution prevention and pollution minimization programs
Review of site environmental history, including identification of past owners of site and activities conducted at the site
Security systems

Compliance Audit
General appearance of facility
Housekeeping practices
On-site, current copy of laws, rules, and regulations
On-site current copy of operation and maintenance manual for systems, processes, operations, and equipment
On-site copy of emergency response plan
Sources of emissions from systems, processes, and operations
Abatement equipment operation and maintenance
Fugitive emission sources at the facility
Visible emissions
Odors
Spills
Emergency response plan and capacity
Check records for operation and maintenance upsets
Appearance of vegetation at and near facility
Sampling of soils, air, water, vegetation, and materials on site
Check for presence of pits, ponds, lagoons, or surface landfills (closed, active, or inactive) as possible sources of air emissions
Check treatment, storage, and disposal of abatement process wastes

example, a check mark accompanied by a reference number would denote completion of the audit of an item and availability of documentation. Another method might consist of recording the results of the audit directly on the form (Table 9-1). Also, the form is flexible enough to permit additional items to be added to each category should the need arise.

Since the administrative phase of the auditing process involves primarily the review of records, many auditors will send a preaudit checklist to the company. This not only expedites the process, but also provides the company and the auditor with a quality control check of the company's attention to compliance and the status of the permits.

As shown in Table 9-1, the auditor must first determine what permits and applications the company legally holds. This is done by checking the copies of air permits and applications on file at the company and then verifying the permit and application numbers with the appropriate regulatory agency. The inventory of existing permits can alert the auditor to requirements in one permit that trigger the need for a permit in another area of regulation. An example of this would be VOC emissions from a water treatment process that would require the submission of an air permit application. The types of permits (i.e., water, air, or hazardous waste) will also govern the extent of the audit of operations. RCRA permitted operations may have to be inspected to determine if there are any emissions regulated under the 1990 CAA Amendments.

The next review should be done to determine the status of the permits that the company currently holds, permit applications that have been submitted and where they are in the review process, and permits renewal dates. With multiple permits and applications, this may become rather complex; therefore, both the company and auditor may find the use of a time-task chart or time-line to be useful in tracking the status of each. This also allows the auditor and company to quickly and clearly see how the air permits and permit applications stand in relation to the other permits. The technical accuracy and completeness of the air permit should be thoroughly reviewed. Finally, the identity of the permit holder should be verified.

The permits should be examined to determine what emissions are regulated. This would include any of the following substances:

- The National Ambient Air Quality Standards (NAAQS) for O_3, CO, SO_2, NO_2, Pb, and TSP/PM-10;
- The National Emission Standards for Hazardous Air Pollutants (NESHAPs) for asbestos, benzene, beryllium, coke oven emissions, inorganic arsenic, mercury, radionuclides, and vinyl chloride;
- Volatile organic compounds (VOCs), as defined by the 1990 CAA Amendments;

- Hazardous air pollutants (HAPs), as defined by Title III of the 1990 CAA Amendments; and
- Air pollutants addressed by state or local agencies.

An inventory of the sources and volumes of emissions should be done next. This includes all emissions from abatement equipment, fugitive emissions, and emissions from any part of a system, process, or operations. The inventories of the emission volumes should be examined to determine if these were derived from calculations or actual measurements. This is particularly important, since the data and information from these inventories will probably be used by the regulatory agencies to set permit fees. Higher permit fees can be imposed if the emissions tonnages are determined based on information from such references as AP-42 or Synthetic Organic Chemical Manufacturing Industry (SOCMI), since these values are usually higher than those obtained using actual measurements. These inventories also set the baseline by which the effectiveness of facility pollution prevention and minimization programs can be measured. Inventories can also be used to identify process and system problems.

The permits should be examined to determine the reporting requirements. This also includes the self-reporting requirements, and the reporting of operation and maintenance upsets and sudden, unplanned, uncontrolled events under CERCLA, SARA, and Title III. The audit should also be done for required record keeping, such as logs of systems, processes, and equipment operation, records for monitoring, sampling, and testing, and compliance records for notices of violation (NOVI), fines, fees, and agency site inspections. Thorough attention should be given to the accuracy and completeness of all records. This should also include a review of the system used to store and retrieve records, data, and information.

Each step of the reporting system should be clearly identified and verified with the company staff. The charting or graphing of the reporting system and process can be of great assistance in identifying areas of deficiency as well as strength in the reporting process. It also gives both the company and auditor a clear picture of exactly how the reporting process works.

The compliance schedule as well as compliance record should be thoroughly reviewed. NOVI fines and administrative penalties should be used to identify problems or deficiencies in facility operations. The compliance record also serves as a baseline of performance and can be used to make those corrections required to meet the terms of the permit. Careful attention to the compliance schedule and record can mean a reduction of operating costs in terms of fines and penalties, plus a reduction in time and energy lost in solving compliance problems. A good compliance record can also be used to promote good community relations. Records should be checked for operation and maintenance upsets. These may be regulated under the terms of the permit and should be documented thoroughly for time of occurrence, duration of the event, measures used to control the event, time of the report of the event, and agency receiving the report. Violations for failure to report upsets should also be noted. A record of consistent upsets in a particular process or system indicates improper operation or maintenance. This should be followed up in the site audit.

The data and information from monitoring programs provide the baseline against which facility operations can be measured. They also can be used not only to confirm the extent and time of exceedances in permit limits, but in defense against wrongful allegations of pollution. This will be most important when pollution from other area sources is attributed to the facility operations. Since this may give rise to legal and regulatory problems, the quality assurance (QA) programs for monitoring, testing, sampling, laboratory analysis, equipment calibration, and operation should be rigorously examined.

The audit of the modeling process should include the identity of the individual or group that did the modeling, the type of model used, the agency approval for use of the model, and the results of the modeling. The specific modeling requirements in the permit should be clearly stated in the audit. This becomes particularly important if more than one type of model is required by the agency to meet the terms of the permit. An example of this would be the requirement for the use of a diffusion model to estimate area HAP ambient air concentrations and a risk assessment model to estimate health effects that would result from exposures to those estimated levels. The appropriateness of the model should also be reviewed. Data and information collected in the monitoring program can often be used to verify the results of the modeling process.

Careful documentation should be maintained for training, safety, and worker right-to-know programs. The times, numbers, and names of employees, the identity of the program instructors, and the program content should be recorded. Not only does this information confirm compliance with existing regulations, but it also can be used in liability and insurance negotiations. This is also a good time to review worker and environmental health records, since both will probably be kept in the safety department files. Excessive sick-leave days, accident reports, and on-site treatment of minor ailments can be indicative of adverse health effects caused by exposures to substances emitted, discharged, or stored at the facility. Suspected sources should be thoroughly checked in the compliance review, and corrective actions should be documented.

Any references to nearby community perceptions of the effects of facility sources on health in the area should be reviewed thoroughly. Regulatory agencies tend to view such reports as potential permit violations and will usually begin an immediate inspection of the facility for sources of such exposures. The prudent auditor and facility manager will not only document but will investigate community health effects complaints.

The facility Emergency Response Plan should be reviewed for completeness, accuracy, and compliance with the permit conditions. It should also be checked to determine if it meets the existing requirements of local, state, and federal air quality regulatory agencies and agency offices of emergency management. Training and education programs to familiarize the staff and employees with the contents of the emergency response plan should also be reviewed.

Pollution prevention and minimization programs and operations and evidence of their effectiveness should be thoroughly documented. Operations and systems where these programs have been initiated should be identified. All signatures on manifests and other compliance documents should be verified. A careful audit of the results of these programs can verify reductions in emissions and lead to the reduction of permit fees and waste and inefficiency in operations.

One of the final items in the Administrative Audit section of Table 9-1 is a careful review of the environmental history of the facility. This includes identifying past owners and operators, all activities formerly conducted at the site, and previous permits issued for operations at the site. This information can be used to establish baseline data and information for the site and in permit negotiations. A careful examination of the facility's environmental history is critical in property transactions. It can serve as a rapid screening method for the detection of abandoned sources of emissions, particularly those of VOCs and HAPs. Such information gains added importance when issues of liability resulting from past practices arise.

Throughout the administrative audit, strict attention should be given to security system strengths and weaknesses. Although a permit may be a matter of public record, that does not mean that all data and information used in preparing the permit application are available for public inspection. Provisions may be written into the permit to protect trade secret information. Both the design documentation and quality assurance audits of the security systems employed to protect both records and the facility itself should be examined. They should be checked for performance, reliability, and vulnerability to external access by unauthorized individuals. Documenting the maintenance of proper security systems can be of great assistance in assuring permit compliance.

A thorough and careful administrative audit can greatly reduce the time required to conduct the compliance part of the facility audit. The results obtained in this part of the audit can be used to rapidly identify problem areas and points of noncompliance. Prompt correction of these problems can result in the reduction of the potential for penalties and fines levied for improper record keeping and practices. This in turn assists in compliance with permit provisions and the reduction of liability.

COMPLIANCE AUDIT

A general compliance audit consists of a thorough and careful inspection of the entire facility. For a special compliance audit, however, examinations of only a specific process or operation may be necessary. Both types of audits, however, are based on the findings of the administrative audit. In either case, the auditor actually examines the condition of the site, work practices, and operations conducted at the facility. While on site, the auditor will check, as a minimum, the items listed in Table 9-1. This phase of the audit may take only a day or over a week, depending on the scope of the audit.

The general appearance of the facility is the first item that should be checked under the Compliance Audit category. Visible emissions, rust-coated structures, unpaved areas, and dust from waste piles are all indicative of potential air permit problems and violations. This is also the image that is presented to the community and regulatory agencies.

Housekeeping practices should be thoroughly noted, since they are indicative of the attention and care given to operations. Littered areas, unswept floors, overflowing trash receptacles, dirty restrooms, and rust and peeling paint are indicative of uncaring management and lax supervision. The facility appearance is also reflected in the quality and manner of employee performance and practice. Poorly trained, indifferent employees are walking permit violations waiting to happen. Areas where poor housekeeping practices and lower employee performance are observed should be specially marked for additional review for permit violations. The location of unpaved areas and uncovered, unprotected waste piles should be carefully recorded. These may be sources of increased fugitive emissions that can lead to permit violations.

The facility should be checked thoroughly to verify that current copies of the following are available on site:

- Applicable laws, rules, and regulations (state and federal);
- Operation and maintenance manual for systems, processes, operations, and equipment; and
- Emergency response plan.

It is never safe to assume that current copies will be available or that employees will know where they can be found when needed. It is most important that their on-site presence be checked in the audit. Employees and staff should be interviewed to determine if they know the location of all regulations, manuals, and plans. A copy of all permits should also be available for easy access by the company staff and auditor.

Continuous and periodic emission sources identified in the administrative audit should be verified during the site inspection. Verification can be done visually or by sampling or monitoring. These emission sources should be checked to determine if they are allowable under the terms of the permit and if appropriate monitoring is being done. These emission sources include:

- Operating parameters data designated in the permit;
- Abatement equipment;
- Any point in an operation, process, or system where a major emission occurs;
- Operation, process, and system points, such as flanges, joints, and valves where small emissions or leaks can occur;
- Open containers and tanks; and
- Open pits, ponds, lagoons, and surface impoundments.

Actual test data are important because they also can be used to prove compliance. Visible emissions, for example, may not constitute permit violations; however, they should be documented thoroughly as proof of compliance. The time of occurrence, duration, physical appearance, and behavior should be recorded. The log for the notification of upsets should be checked for accuracy and completeness in reporting the appropriate information to the required regulatory agency.

The presence of odors at the facility should be noted with respect to type of odor, duration, association with facility operations, and community response. Odors are important indicators of process changes and system fluctuations. While odors may not be addressed in the permit, they can serve as excellent sentinels of impending or recent violations of emissions limits set forth in the permit. Most state and local air quality control programs prohibit malodors at the facility fenceline. Although this regulation is loosely enforced, it can be used by agencies as an additional means of imposing penalties and fines. If the odor persists, ambient air sampling should be done and the analysis results evaluated in terms of potential odor sources located at the facility. Odors can be the first indicators of an "invisible" spill. This is a spill that involves material that evaporates quickly and leaves little or no trace of the initial spill.

Since spills may or may not constitute a permit violation, any visible spill should be documented with respect to fulfillment of reporting requirements, time and duration of event, material(s) and volumes involved, and containment and remedial actions taken. The facility emergency response plan should contain a section for spill control and containment. Procedures for spill containment should be posted in the workplace, and spill response training and exercises should be documented. The auditor should check to see if spill control equipment and materials are available and in proper operating condition. Employee spill response training should also be documented.

The Emergency Response Plan (the Plan) of the facility should be verified by the auditor by checking the physical resources available for implementing the Plan. Employee emergency response training should be verified. Community resources, such as the fire department and police, should be contacted either verbally or in writing to ascertain their availability to render assistance during an emergency. Also, community awareness of the existence of a facility emergency plan should be noted. In addition to protecting health, emergency preparedness can be used to prevent permit violations and reduce liability.

The appearance of vegetation on site or near the facility is a good indicator of exposures to hazardous or toxic substances. If visible areas of vegetation damage exist, the person in charge of the audit may elect to have a specialist, such as a botanist or zoologist, conduct a special audit of the damage. Every effort should be made to determine the source of the exposure that caused the damage. If the damage was caused by emissions from a source on site, this could lead to one or more permit violations. Areas of vegetation damage should be documented in detail and plotted on a facility map. Samples should be taken according to a specified sampling protocol and stored under chain-of-evidence procedures.

Samples of soils, air, water, and vegetation should be taken on site and analyzed as a part of establishing baselines and information for those substances listed in the permit. Other materials stored on site should also be sampled as a part of the audit process.

From information obtained in the administrative audit, a thorough inspection for emissions should be made of all pits, ponds, lagoons, and surface impoundments, both active and inactive. Monitoring data and information should also be examined to detect the presence of emissions from these sources. Any additional inspection and monitoring provisions for VOCs and HAPs that might have been written into the permit should be included in the audit. Comparisons of the present and past visual inspections and audit documentation should be made if at all possible.

The last item included in the Compliance Audit section of Table 9-1 is that of the audit of the treatment, storage, and disposal of abatement process wastes. These wastes may be classified as hazardous waste or may contain other regulated materials, such as VOCs, PCBs, asbestos, or HAPs. If testing identifies these abatement process wastes as a hazardous waste, special TSD procedures will have to be followed and an RCRA audit done.

The compliance audit consists of making on-site visual observations and documenting these observations. All employees should assist in the audit whenever possible. Long-term employees of the company can be invaluable sources of information regarding changes in procedures and operations at the facility. This phase of the audit also involves sampling and monitoring soil, air, water, and waste, as well as substances and mixtures produced at the facility or used in operations and maintenance. All of this should be done in a spirit of cooperation between the auditor and the company. Unless this can be achieved, the success of the audit will be limited. The audit should be used to assist the facility manager in reducing air emissions from the facility and in achieving the level of compliance set forth in the permit.

AIR PERMITTING UNDER THE 1990 CAA AMENDMENTS

The comment most frequently heard in the environmental community today is that the 1990 CAA Amendments are the most complex and comprehensive environmental legislation passed to date. This is nowhere more evident than in the permitting process. It is comprehensive in that all permits must be reissued and is complex in that permits are issued in terms of both types of emissions and permitting programs. Previously, permits were health based and issued on the basis of "end-of-the-pipe" control of emissions from a facility. The 1990 CAA Amendments will not only require this type of control, but are considered to be "enforcement based."

In addition to having permits written and issued in the context of achieving these objectives, the 1990 CAA Amendments also include new categories of pollutants for air toxics or hazardous air pollutants (HAPs), volatile organic compounds (VOC), acid rain, and stratospheric ozone. The scope of the 1990 CAA Amendments has set forth a significant number of changes in the permitting process. Table 9-2 provides a summary of these in a listed form that can also be used as a screening checklist to determine which programs and pollutants will trigger the permit application process. This is essentially an audit of the new regulatory requirements for air permitting. Using Table 9-2 as a checklist, the company and auditor can rapidly determine under what programs and pollutants the permit will be regulated.

Table 9-3 shows how the pollutants and programs regulated under existing permit conditions can be compared to the changes that will be required in preparing the new permit application under the 1990 CAA Amendments. This is essentially a matrix in which the conditions of the permit issued under the 1977 CAA Amendments serve as a baseline from which to evaluate the direction and magnitude of changes required by the 1990 CAA Amendments. As with the General Audit Form (Table 9-1), these changes can be entered

TABLE 9-2 Summary of Major Provisions of the 1990 CAA Amendments

Establishment of a federally enforceable permitting program.

States may implement their own programs if approved by the EPA by 1993.

Facilities must file new permit applications by 1995.

Issuing permits for a fixed term not to exceed 5 years.

Notice of permit applications to be given to states within 50 miles of the source or with an affected air quality area.

Expansion of regulated facilities to include small businesses that:

- Employ < 100 individuals;
- Do not emit > 50 tpy of any regulated pollutants;
- Emit < 75 tpy of all regulated pollutants;
- Are not a major stationary source; and
- Defined in the Small Business Act.

Imposition of a permit fee of $25/ton of emissions.

Sources subject to the PSD permit program include:

- Any one of sources in 28 listed categories with 100 tpy of potential emissions of a regulated pollutant;
- A new source with 250 tpy potential emissions of a regulated pollutant; and
- Modification of an existing source of regulated pollutants exceeding listed tpy emissions.

Monitoring and meteorological data required for permit.

Increase in regulated pollutants to include volatile organic compounds (VOC), and hazardous air pollutants (HAPs).

Regulation of pollutants under the Acid Rain Program (Title IV) and Stratospheric Ozone Depletion (Title VI).

Permits must contain best available control technology (BACT) for analysis of emission control devices.

Air quality–related values analysis required.

Netting out requirements for the construction of new modification of major sources (reductions of emissions within a facility equal to the new source emissions).

Offset emissions requirements for new source review (NSR) (between facilities, it allows emission reductions equal to new source emissions).

Provisions for increased public participation in the permitting process.

Air toxic program for permits will include:

- Sources emitting 10 tpy of listed substances;
- Sources emitting 25 tpy of combined listed substance;
- Application of maximum achievable control technology (MACT) for new sources;
- 90% reduction from the emission source 1987 baseline level;
- > 750 categories and subcategories of industrial sources; and
- Offsets or netting out.

Acid rain program will include the following provisions that will affect the permitting process:

- 10 million tpy SO_2 reduction from 1980 levels;
- Allowance system for SO_2 based on 1985–87 average fuel used, sold, banked, consumed, or traded;
- 2 million tpy NO_x reduction from 1980 levels;
- Mandatory continuous air monitoring; and
- Fees of $2,000/ton for excess emissions.

TABLE 9-2 (*continued*)

Preconstruction permits required under PSD and NSR.

Construction permits required for sources having existing state air permits, PSD, NSR, and HAPs.

Renewable permits required for sources having existing operating permits and sources subject to CTG.

Operating permits required:

- NSR.
- Any major source.
- Any source subject to air toxics provisions (Title III).
- Any source subject to acid rain provisions (Title IV).
- Any source presently permitted by the state.
- Any source subject to NSPS.
- Permit requirements will include emission rates, certification of compliance, compliance plan, record keeping, and self-reporting.
- Veto by the EPA of any permit.
- Sources emitting 100 tpy in any attainment area.
- Emission limits for non-attainment areas:

moderate:	100 tpy CO
	100 tpy PM-10
	100 tpy VOC & NO_x
	50 tpy CO
serious	70 tpy PM-10
	50 tpy VOC & NO_x
severe:	25 tpy VOC & NO_x
extreme:	10 tpy VOC & NO_x

directly on the form in numerical form or by using a notation accompanied by a reference for detailed information. The general audit form (Table 9-1) for air permits can still be used in the compliance audit of the facility. The state authority in permitting has been included as a separate category in Table 9-3 because states can impose more stringent air quality control requirements and many will. It may be necessary to expand the compliance audit (Table 9-1) to meet the new regulations; however, it is sufficiently flexible to allow such changes to be easily incorporated.

The scope of the new regulations will greatly expand the need for auditing in each step of the permit process. The first step in the process should be to conduct a regulatory audit, using the checklist shown in Table 9-2. The next step should be to perform an audit of the terms set forth in the existing permit. The information obtained in these two audits can then be compared, as shown in Table 9-3, to determine what new pollutants and programs will have to be included in the new permit application. Once this has been determined, the

TABLE 9-3 An Audit of Permit Requirements Under the 1977 and 1990 CAA Amendments

		Emissions									
		NAAQS									
	CAA Permit	SO_2	CO	NO_x	Ozone	Pb	TSP	PM-10	NESHAPS	VOC	HAPs (190)
Permits other than air permits											
State Authority	1977										
	1990										
SIP	1977										
	1990										
PSD	1977										
	1990										
NSPS	1977										
	1990										
Preconstruction	1977										
	1990										
Construction	1977										
	1990										
Operations	1977										
	1990										
Non-attainment areas	1977										
	1990										
Ozone depletion	1977										
	1990										
Acid rain	1977										
	1990										
Health and risk assessment	1977										
	1990										

new permit application can be written, reviewed, and submitted. The information obtained in the first and second steps of the permit application audit process can be used in the review to check the accuracy and completeness of the application.

Although the publication of all the final regulations for the 1990 CAA Amendments is not anticipated before 1995, there is still much that can be done to prepare for the new permitting application. Prudent companies will begin a review of the provisions of the 1990 CAA Amendments to determine what major changes this will require in existing permits. Certainly, companies will want to include long-term planning for capital investments in terms of how these changes will affect the construction of new facilities and modifications of existing facilities. All of this will call for the development of permit auditing strategies. The most effective of these new permitting and auditing strategies will be based on the baselines established by today's auditing practices.

References

Baumer, A. R. 1982. Making environmental audits. *Chemical Engineering*, November 1, pp. 101–104.

Capasso, E. 1986. Technical considerations in conducting a hazardous waste facility audit. *Proceedings of the Haz-Pro Conference*. Northbrook, Ill.: Pudvan Publishing.

Cheremisinoff, P. N., and J. Ten Eyck. 1987. Environmental auditing: a basic guide. *Pollution Engineering*, April, pp. 72–75.

Clean Air Act Amendments of 1990, PL 101-549, November 15, 1992.

EEI. 1983. *Environmental Auditing Workbook*. Edison Electric Institute.

Environmental Protection Agency. 1985. Environmental auditing policy statement. *Federal Register*, Vol. 50, No. 217, November 8, pp. 46504–46508.

Guida, J. F. 1982. A practical look at environmental audits. *Journal of the Air Pollution Control Association*, Vol. 32, No. 5, May, pp. 568–573.

Hale and Dorr. 1991. *The Clean Air Act Amendments: Strategies for the 1990s.* Washington, D.C.: Hale and Dorr.

Kennedy, B. 1982. The need for environmental audits. *Pollution Engineering*, July, pp. 28–30.

Laznow, J., and S. H. Kopp. 1992. Developing corporate compliance strateies for HAP emissions. *HAZMAT World*, March, pp. 53–55.

Sokolik, S. L., and D. J. Schaeffer. 1986. Environmental audit: III. improving the management of environmental information for toxic substances. *Environmental Management*, Vol. 10, No. 3, pp. 311–317.

Splendore, J. L. 1983. The how and why of environmental audits. *Industrial Wastes*, March/April, pp. 14–15.

Truitt, T. H., et al. 1983 *Environmental Audit Handbook*. New York: Executive Enterprises Publications Co.

10

Air Emissions, Environmental Auditing, and Waste Minimization

Paul D. Rosenstock and Sam N. Popowcer

THE REGULATORY ENVIRONMENT

Many people think that environmental regulations are a recent phenomenon—a necessary answer to increased environmental pollution that has resulted from the rapid urbanization and industrialization of modern society. This is not so. "The first person recorded to have suffered from pollution was a Queen of England, Eleanor, who was driven from Nottingham Castle in 1257 by unpleasant fumes from sea coal (high in sulfur) burned in the city below. By the last decades of the thirteenth century, London had the sad privilege of becoming the first city in the world to suffer man-pollution. And in 1307 a royal proclamation was made in Southward, Wapping, and East Smithfield forbidding the use of sea coal in kilns under penalty of forfeiture." (Gimpel 1976)

Environmental regulations in the United States have progressed to the point where almost every aspect of chemicals manufacture and use is covered by some law or regulation at the federal, state, or local level. Initially, the federal government was the prime mover, but recently some states, notably California, New Jersey, and Texas, have outpaced the federal regulations. In this chapter, our discussions will be limited to the federal laws and regulations that are enforced by the Environmental Protection Agency (EPA). The "regulatory wheel" shown in Figure 10-1 lists some of the major laws enacted to implement the National Environmental Policy Act of 1969, which set out the priorities for the federal environmental program.

This chapter will address the 1990 Clean Air Act (CAA) Amendments as the most recent and prime example of regulations that mandate extreme reductions in environmental pollution. We will discuss the procedures for and format of environmental audits designed to monitor compliance with envi-

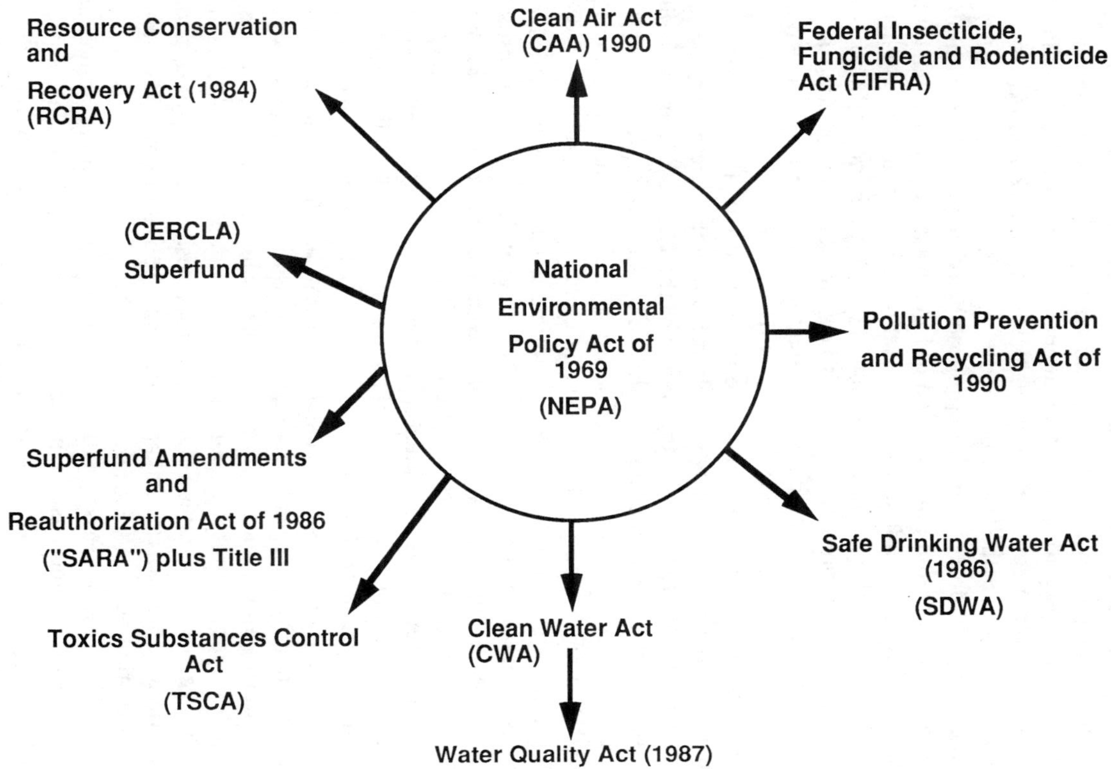

FIGURE 10-1. The regulatory wheel.

ronmental regulations and to measure the extent of past polluting practices. We will also provide examples of operating and management practices that can be used to minimize the potential for environmental pollution. In reality, the goals of environmental compliance are twofold: to reduce or eliminate the amount and type of waste materials (pollutants) that can adversely impact the environment and to monitor, document, and report these reduction activities in a manner consistent with regulatory mandates. A third aspect of compliance is that it is the law of the land!

The goal of the most current and proposed environmental regulations is to reduce environmental pollution and to force waste minimization, either through the use of mandated technology or through extensive audits of processes with the subsequent establishment of waste minimization goals. In addition to the regulations, the cost of waste generation—both disposal and long-term liability—has forced industry to adopt the religion of waste minimization as necessary to remaining competitive.

Waste can be defined as any material emission to the environment, whether from the manufacturing process or from support activities; it is any discharge to the air, the land, or the waters of the United States. In reality, waste is as anything that does not get utilized by the ultimate customer. If a process generates a waste, pollution occurs. Pollution can come from the required disposal of nonsaleable waste materials generated by a process. It can come from non-process–related materials management activities, air emissions from storage units, leakage from storage units, discard of obsolete and off-grade final products, and so on. The whole process or operation must be addressed in a pollution prevention program.

Until very recently, the individual regulations addressed a single media—air, water, or solid waste. Thus, much of the pollution abatement efforts were actually media transfer of pollutants. For example, abatement of air emissions with a scrubber generates contaminated wastewater that requires further treatment, either by biological treatment or solids removal. Biological treatment of wastewater removes (destroys) the pollutants, but results in a biosludge that requires disposal, either by incineration or in a landfill. Similarly, solids removal generates a sludge that requires treatment. Most media transfer was acceptable as long as cost-effective disposal sites were available. But this situation no longer exists.

As noted above, it is necessary to properly monitor, document, and report pollution prevention and waste minimization efforts in order to demonstrate regulatory compliance. To assure yourself that the required systems are in place, periodic environmental audits should be conducted. The environmental compliance audit is used to determine the status of current environmental management and document programs in place, to collect data on progress and implementation of programs for further compliance requirements, and to evaluate the extent of environmental problems at a particular site that have resulted from past practices. The outline of this process is discussed below.

At first glance, air emissions, environmental auditing, and waste minimization would appear to be diverse topics out of the entire regulatory compliance matrix. But as we have discussed, they are closely related and provide the complete flow of regulatory compliance. We hope to answer the following questions in this chapter:

- Air emissions—What do we have to do to comply with the 1990 CAA Amendments?
- Environmental auditing—How do we determine environmental compliance?
- Waste Minimization—How can we improve our environmental compliance?

Finally, we have included a case history drawn from our own experiences in these areas, to demonstrate some of the process changes that can be done to meet regulatory demands. Also note that pollution prevention saves money, either by utilizing resources better or by reducing long-term liability.

THE 1990 CLEAN AIR ACT AMENDMENTS

The 1990 CAA Amendments are the culmination of two decades of environmental regulations. They are complex, address many issues, and set specific timetables for compliance. They also have very broad coverage that will bring many currently unpermitted air sources into a permitting program. For the chemical, pharmaceutical, refining, and allied industries—indeed, for all chemical users—Titles I, III, V, and VII are the key sections of the act. Title I addresses non-attainment (the failure to meet federal ambient air guidelines for ozone, oxides of nitrogen, carbon monoxide, particulate matter 10 microns or below (PM-10), sulfur dioxide, etc.). Title III addresses hazardous air pollutant (HAP) emissions, and specifically names 189 chemicals and chemical classes that are covered by this title. Title V lays out a very ambitious federal permitting program for all major sources, and Title VII covers enforcement and defines new criminal penalties. The reader who is interested in a more in-depth understanding of the 1990 CAA Amendments should read the act (over 700 pages of fine print) and/or read one of the many books or journal articles that cover this subject.

This said, there are some specific points that must be made. The 1990 CAA Amendments require the EPA to develop source categories and subcategories and then to mandate technology for the control of air pollutants from major sources in each category. "Major sources" is defined in both Title I and Title III. For ozone non-attainment, a major source varies from 10 to 100 tons per year (from 2.3 to 23 pounds per hour) of volatile organic chemicals (VOC) or oxides of nitrogen (NO_x) emissions, depending on the attainment classification of the area. A major source under Title III is any source that emits 10 tons per year of a single HAP or 25 tons per year of two or more HAPs.

The 1990 CAA Amendments requirements will come into play over the next 10 to 15 years, but you can expect most industrial categories to be impacted over the next 7 years. Timetables have slipped due to limited resources, legal challenges, and so on, but the regulations will march onward toward their inexorable goal of minimizing the discharge of pollutants to ambient air. The 1990 CAA Amendments should not be looked at as merely an air program; rather, it is at heart a source reduction program. If a process does not use offending material, it will not emit them, and, if a process uses them, process changes and abatement devices must be used to minimize emissions to the extent technologically possible. Transfer of pollutants from one media to another, as is done with scrubbers, is no longer acceptable for general use. This does not mean that scrubbers will disappear. They are still the abatement device of choice for removing acid gases from air streams.

The heart of the 1990 CAA Amendment is the permitting program that will require very detailed permits on many thousands of sources that may not be permitted under current regulations. Succinctly put, the 1990 CAA Amendments permitting program is an National Pollutant Discharge Elimination System (NPDES) permit for air, in an RCRA Part B Permit format, wrapped in a pollution prevention goal. The impact of the 1990 CAA Amendments will be profound, as it will touch every aspect of manufacturing—even manufacturing that is now on the fringes of or exempted from the regulatory process, such as factory bakeries. The EPA is mandated to develop Control Technology Guidelines (CTGs) to define technologies for the reasonably available control technology (RACT) required for Title I sources. Maximum achievable control technology (MACT) standards must be set for each category and subcategory and will be required on all major sources that fall under Title III. New sources will have to achieve the lowest achievable emissions rate (LAER).

The Title V permit will be very confining and will require considerable documentation. Table 10-1 lists the major problems that manufacturers and chemical users will be faced with. These constraints are:

1. *Decreased operating flexibility* will result from a permitting system that will require all allowable operation (raw materials, processes, emissions) to be explicitly stated in the permit. Any deviation will require agency approval or a permit modification. Processing a new or amended permit will of itself decrease your operating flexibility because the process can be painfully slow.

TABLE 10-1 Major Effects of the Clean Air Act Amendments

Decreased operating flexibility
Increased monitoring requirements
Increased records keeping
Increased public and agency oversight
Increased operating costs
Increased civil and criminal liability

2. *Increased monitoring* will be required both to certify compliance with your permit and to satisfy yourself that you are meeting your permit. The authors believe that continuous emissions monitoring for HAPs will be mandated as the technology to measure these emissions develops. Fourier transform infrared spectroscopy (FTIR) is a likely candidate technology.
3. *Increased record keeping* will be required for emissions monitoring, equipment operating conditions, repair and maintenance, leak detection and repair programs, operator training, and so on. Record keeping is hard to maintain and thus provides an easy way for inspectors to find violations.
4. *Increased agency and public oversight* will call more attention to your operations. We can state from first-hand experience that a chemical manufacturer or chemical user cannot ignore public concerns. It also takes a lot of senior personnel time to address the community issues. All of these added operational, record keeping, and equipment and process changes—all part of compliance with the 1990 CAA Amendments—will increase operating costs! Another problem will be the increased civil and criminal liability that the 1990 CAA Amendments now imposes. In addition to operating and emergency procedures, an audit program to assure that procedures are being followed will be needed to demonstrate compliance.

The key to compliance with the 1990 CAA Amendments, as it is with any regulations, is to have accurate information about your processes, to do up-front planning to develop a compliance strategy, and to participate in the regulatory process.

An audit should be conducted now that can be developed with baseline data on each process. Since the bottom line is to decrease and/or eliminate emissions, both from point and fugitive sources, it is extremely important to determine what the process emissions are by actual measurement. Unfortunately, many of the SARA 313 numbers were developed using calculations that may or may not be correct. A Chemical Manufacturers Association (CMA) sponsored study showed that the EPA SOCMI emission factors over estimate fugitive emissions by a factor of 10 to 100 (BCM Engineers 1989). Succinctly put, to operate under the 1990 CAA Amendments, both the chemicals and the quantity that are emitted from each process and plant must be known.

Emissions monitoring data can be used to determine which processes use and/or emit air toxics and/or VOCs. Monitoring also provides the data needed to determine which sources are covered by the 1990 CAA Amendments. It has been our experience that many companies are not meeting their current air permit emissions limits or do not have permits for existing sources that require permits. Everyone should get their house in order as soon as possible. Examine permits to make certain that the operations are in compliance and to determine if additional permits are required under existing state regulations. The civil and criminal penalties of the 1990 CAA Amendments make it imperative that all existing sources be brought into compliance with existing regulations as soon as possible.

We believe that a mini–health risk assessment should be done based on the measured emissions, to determine if any potential ground-level toxicity problems exist. The EPA will develop a program for risk analysis by the year 2000, but several acceptable methodologies for health risk assessments now exist. Health risk, real or imagined, will be the major focus of citizen suits, and potential problems should be addressed and remedied now. Finally, the permit application will require a listing of compliance history. Everyone will want to present the best compliance history possible and thus should quickly correct any problems that exist.

Completion of the emissions audit should provide the information needed to develop a strategic plan for compliance. As already noted, this plan will have to be site specific. A determination of the categories and subcategories of the operations will provide a better idea of the timetable for MACT requirements. It is also very important to realize that Title I and Title III have different technology requirements. An air toxic can be a VOC, but may be regulated under only one Title, depending upon the amount emitted and the attainment classification of the area in which your plant is located. You will have to develop a strategy to meet the requirements of both titles.

You should develop an ongoing emissions inventory program to make certain that your emissions data are as accurate as possible. This is especially true for batch operations that will handle numerous raw materials over the period of a year. In addition, you should start to collect the data needed for permitting so that you have sufficient time to develop data that are not readily available. Finally, you should amalgamate all of the data and develop a permitting strategy. You should determine the costs of various options, what process changes can be made, and when they should be made. Voluntary early reductions should also be considered.

An effective communications program is also an absolute necessity, since the permitting process has public input and the possibility of public hearings.

You cannot control information flow unless you have an in-place communications program. The CMA Responsible CAER program is a good guide to use in setting up a public information program. Communications should not be limited to outside your fence line. Employees are the most valuable ambassadors to the community and should be kept well informed. In addition, we have observed that any program involving changes in work practices will not get off the ground without active and willing support by the personnel who will have to implement the change. Changes in work practices will surely be required to reduce fugitive emissions, spills, and so on, and employee cooperation will be imperative.

Finally, industry should make a maximum effort to influence the regulatory process. Membership and active participation in trade and business organizations is an excellent way to participate in the process. Since most of the 1990 CAA Amendments action will be at the state level, you should also establish lines of communications with your state legislators. Many regulations are reviewed and/or written with the help of advisory panels, and participation on a panel is an excellent way to influence the regulatory process. Finally, comments on key laws and regulations should be made to make certain that the specific issues of concern to your operations are addressed.

In conclusion, remember that compliance with the 1990 CAA Amendments regulations will be difficult and costly. You can and should exert a measure of control over your destiny by setting up a compliance program now that will enable you to determine the most cost-effective options for compliance. Companies that do up-front planning will be in a much better position to maintain a competitive edge.

THE ENVIRONMENTAL AUDIT

General Environmental auditing is the process by which a complete and in-depth review of current and past regulatory and pollution control practices is done for a facility or process. The audit looks at discharges, emissions, raw materials, intermediates, and products that could have impacted the environment through present or past practices. The need for environmental audits is obvious in light of the demands of many regulatory compliance programs.

- The Pollution Prevention Act of 1990 requires the establishment of specific source reduction/waste minimization goals as part of a Pollution Prevention Plan. If these goals are not achieved, environmental audits may be mandated. In addition, current and projected solid waste regulations require the preparation of source reduction strategies, which implicitly require full operation and management review.

- There is an obvious need to review ongoing operations to ensure compliance with existing regulations related to material management, public safety, employee safety, fire, explosion, and so on. Such a review is directed toward ensuring that the operation is not causing imminent or potential harm to employees or the public, as well as addressing concerns about the public perceptions of overall corporate compliance activities.
- The requirement for a complete environmental audit or assessment of contamination potential may be specifically mandated or implicitly directed by federal or state regulations that address aspects of property transfers. Examples include superlien and lien laws, such as Massachusetts General Law Chapter 21E, that place liability for clean up of hazardous materials on property owners, regardless of fault. As a result, lending institutions in Massachusetts usually require that a property have a site assessment done before they will lend the money to purchase it. There are also Environmental Responsibility and Clean-up type laws, patterned after New Jersey's Environmental Cleanup Responsibility Act, (ECRA) which mandates site assessments at certain types of "facilities," as triggered by certain types of "transactions."

There are several reasons why a company might choose not to conduct an environmental compliance and/or liability audit. These reasons could include the following:

- Fear of "self-incrimination";
- Concern for the confidentiality of the audit findings and the associated potential for damage to corporate and/or individual reputation or public image; and
- The belief that ignorance, or "not knowing," may be to their advantage.

To facility managers, the audit can be perceived as a threat to their reputation. To the company, the audit and the confidentiality of the audit results can be a threat to the corporate image from both a public and regulatory viewpoint. *But there are dangers in "backing off" from an audit.* If a company fails to exercise diligence, particularly in cases where property transfers or public health and safety are concerned, they might be charged with negligence if an environmental problem is discovered, which, from a regulatory viewpoint, should have been identified sooner.

It is for this reason that the ignorance plea, or the "what-you-don't-know-can't-hurt-you" attitude is recognized as a major pitfall associated with the "audit" vs. "no-action" decision-making process. As a result and in accordance with current regulatory thinking, it is recognized that enforcement actions and legal judgements may be weighed on the side of "should they have known?" rather than "did they know?" There is a general consensus that the sooner a company identifies its environmental problems and acts to rectify them, the better off it is.

In summary, we note that, while environmental auditing is not a regulatory mandate or requirement, it is:

1. Transaction driven in the case of environmental impairment.
2. Operating management driven in the case of environmental compliance.
3. The basis in regulation primarily as a means of developing information that may be required to report some environmental condition, control, or documentation status. While not mandated, it may be necessary to ensure and document regulatory compliance and source reduction implementation.

Because of the potential legal issues involved with utilizing information developed during the auditing process, an environmental lawyer is a necessary part of an effective environmental auditing program. The lawyer should be involved in the initial audit planning phases so that the flow of developed information can be appropriately controlled and channeled.

The Audit Direction

The key to compliance with any regulation is knowledge of your process. An excellent way to gain this knowledge is through a properly conducted environmental audit and through appropriate follow-up actions based on the findings of the audit. Much of the information that an audit needs already exists, but it is usually not presented in a format that makes data retrieval easy. It is also common to find that some required information does not exist. A properly conducted environmental audit should provide all of the information required to institute a pollution prevention program and to develop strategic plans for meeting regulations, especially the 1990 CAA Amendments. The CMA's *Pollution Prevention Resource Manual* is typical of the auditing guides that various trade organizations have developed.

As stated above, improper record keeping is the cause of many regulatory inspection citations. Proper records are difficult to maintain, and errors or omissions are easy to spot. Thus, the audit should pay particular attention to record keeping and to the systems that are in place to ensure that the records are properly maintained. Follow-up actions to correct problems should be immediate and should continue until the problems are resolved. Pay particular attention to the retention times mandated for specific documents in the regulations. Also, develop a master list of where all documents required by permits and regulations are kept at a particular site. This listing should become part of the site emergency management plan.

Many regulations require that specific records be maintained. For example, the holder of an RCRA Part B Permit must maintain operator training records, inspection records, an operating log, and manifests. Most air abatement permits require that the operating conditions of the abatement device be recorded (fluid flow for scrubbers, temperature for vent incinerators). Releases to the environment must be reported to the National Response Center under Superfund and the Clean Water Act, and to the Local Emergency Response Commission under SARA. The audit should look at all applicable regulations and ensure that the required procedures are in place and that the minimum required records are kept and properly filed (stored).

Record keeping is important, *but the core of a properly executed environmental audit is the determination of emissions to the environment—to air, water, and land.* The best starting point is SARA 313 reporting data for the site. However, it is very important to note that many companies used EPAs SOCMI factors to estimate air emissions by calculation and that these factors can grossly overestimate emissions (Environmental Protection Agency 1990). Thus, it is very important to use actual measured emissions whenever possible. This is especially true for data to be used for 1990 CAA Amendments permitting because measured emissions will be required for compliance reporting. Very accurate baseline data are also needed to design abatement equipment that will be mandated by the 1990 CAA Amendment for submitting your Title V air permit, and to assure that you can and are meeting your permit conditions.

Fugitive emissions can be measured using approved methodologies, such as EPA Method 21. Point source emissions can also be directly measured using the appropriate EPA or state test methods. Emissions from the storage tanks, which are also point source emissions, can also be measured. Care must be taken to make certain that the data are representative of filling and emptying operations, as well as diurnal changes in temperature. Emissions from lagoons are difficult to measure and should be calculated, with the understanding that the calculations will usually yield high estimates. Ambient air monitoring around lagoons is possible but expensive. It is also a good idea to compare measured emissions with calculated emissions and to investigate any gross discrepancies. Also, measurements on batch processes should be done to define both the "worst case" and the "average" conditions.

Waste streams for disposal or recycle can be analyzed by the Toxic Constituent Leaching Procedure (TCLP) or by analytical methods used for the products. Wastewaters are analyzed by the EPA 600 series Methods or the TCLP method. Analytical data on solid waste should be available from disposal sites and from disposal forms submitted to disposal sites and state agencies for approval. Wastewater analyses are probably required as part of pretreatment contracts with Publicly Owned Treatment Works or as part of the NPDES Permit requirements on-site treatment plants.

Material balances are an excellent way to gain an understanding of a process. But a material balance rarely substitutes for analytical data where emissions are in the part per million and part per billion range. In this era of ever-tightening regulations, it is extremely important to know what the trace contaminants are in raw materials and wastes. For example, a chemical manufacturer was surprised to find ethyl benzene in his wastewater at several hundred parts per billion, because they did not handle ethyl benzene as a raw material in their plant. On investigation, it was determined that the ethyl benzene was a contaminant in styrene, which remained inert during processing, and was transferred to the wastewater during purification steps. In another case, a plant found high ppb levels of toluene and benzene in their wastewater that were ultimately traced to spillages of unleaded gasoline.

An adjunct to the material balance is a mini–risk assessment to determine where any employee or community exposure liabilities may be. This will enable you to set priorities for follow-up work. Since the current emphasis in on "toxic reduction," a waste minimization program should prioritize actions based on toxicity (see below).

The audit should carefully check existing permits to make certain that all requirements are being met. It should also look at all operations to determine if additional permits are required based on existing regulations. It is not uncommon for a facility to have made modifications without realizing that a permit or a permit modification was needed. Further, many air abatement devices were installed under permits, but were never tested to assure that they met permit conditions. It is amazing how many permitted abatement devices do not meet mandated removal efficiencies and/or emissions levels.

Finally, the audit should look at past compliance history. More of the laws and regulations are mandating public input to the permitting process. Since past compliance is of major concern to the public, it is necessary to immediately address all noncompliance issues as soon as they are discovered. You cannot convince the public that you can properly operate your facility if you have a record of continuing compliance problems.

The Audit Format

In order for the environmental audit to be an effective, useful, and economical tool for gathering environmental information, it should follow a specific phased, format (Table 10-2). This format is not based on any set standards and can vary extensively, depending upon the size, type of operations, locations, and so on, of the individual facility or corporate structure being audited.

Table 10-2 The Environmental Audit

Determine emissions from all sources:
- Calculated?
- Measured?

Do a material balance on process.

Review existing permits for compliance:
- Are permit conditions for emissions, removal efficiency, and operating conditions met?
- Is record keeping in order?
- Are additional permits required under existing regulations?

Review past compliance history.

Evaluate health effects of emissions.

Submit written report with findings and recommendations.

There are, however, several general guidelines that should be addressed in setting up the specific program to be adopted.

1. *Commitment/protection*—It is absolutely imperative for corporate management to understand that environmental audits are a useful and necessary tool. They are used to develop information on facility operations and are used to minimize the potential for liabilities and penalties from environmental law violation. As already mentioned, a qualified environmental lawyer should be involved in the auditing process.
2. *Phased approach*—The audit should be set up with specific information and compliance goals in mind. For example, are the program objectives to assess and reduce the cost of regulatory compliance? Or are the goals to assess the current monitoring and compliance status of emission control devices to assure documentation and operation correctness? Are the goals to assess and assure facility management structures in establishing and implementing mandated compliance programs for training, inspection, reporting, follow-up, and so forth?
3. *Data gathering program*—Determine what program and procedures will be followed during the audit. Will data be gathered by questionnaire, by physical facility inspection, or by both (the preferred program)? Will the audit be conducted by in-house personnel or by outside consultant? Will all facilities be audited, or will initial efforts be directed towards "typical" installations? An effective checklist for developing working documents to be used by a facility inspection can be developed from the *Multi-media Compliance Audit Inspection Procedure*, which is available from the EPA Office of Legal and Enforcement Counsel, National Enforcement Investigations Center, Denver, Colorado. These federal guidelines should be reviewed to ensure that inclusion of state and local requirements are appropriate.

4. *Team approval*—Choose personnel who are knowledgeable in the management and operation of the facilities to be audited. The audit should not be a learning experience for audit team leaders. Note, however, that the audit can be used to train future auditors, but that these trainees should not be official members of the audit team. Depending on the size and complexity of a facility being audited, an individual team should have between two and four members. Auditors chosen for the team should be appropriately trained and instructed in the goals and the data gathering mechanism designed for this audit program.
5. *Audit scheduling*—The audit should be scheduled based on such factors as auditor availability and facility operating schedule. Where possible, the actual facility visit should be preceded by submission of a pre-visit questionnaire. The audit team should review the questionnaire response against appropriate audit scope and relevant regulations. The audited facility should be advised, in advance, of the audit protocols that will be followed.
6. *Audit initiation/implementation*—Care should be taken to assure that audit schedules and protocols are followed by the team and by the audited facility. To the extent practical, the audit team must be strong in its insistence that necessary plant personnel and information are provided based upon the preliminary requests and agreed upon programs. During the site visit, the team should meet with facility management before the audit is begun to explain the process and answer questions. There also should be a debriefing meeting between the audit team and site management to discuss the results of the audit. The audit team must avoid excessive preliminary judgements in this debriefing. Also, the data collected during the audit must be reviewed with the appropriate facility personnel to assure that the information accurately reflects facility management and operations practices.
7. *Documentation/reports*—Based upon the facility debriefing and data review, the audit team will prepare an audit report. The audit report should follow a consistent format that has been developed at the time that the data gathering program has been set up. Report consistency between a company's facilities is critical to the audit program, so that all facilities may be compared for common denominations. Audit reports should highlight items of noncompliance or potential future liability as the first issues to be addressed.

   To be a useful management tool, the report must provide specific suggestions or recommendations that address each of the items of noncompliance or potential liability. In some cases, the suggestion or recommendation may be for additional investigation or study, but each item must be addressed. A legal review of the draft report should be done prior to preparing and releasing the final report. The draft report should also be reviewed once again by the audited facility manager. The final report can then be issued to appropriate upper management personnel, but the number of copies made and distributed should be as small as possible.

8. *Remedial action responsibilities*—A successful audit program must follow up to make certain that all noncompliance or liability items are immediately addressed by appropriate personnel. Audit team members or individual facility managers should be assigned to monitor and document remediation activities. The report should not be considered complete until these assignments are made.
9. *Audio follow-up/progress review*—Assigning follow-up action responsibility does not end the audit program. Procedures must be established to insure that remediation activities are being implemented. Depending on the capital expenditure requirements of the remediation activity, some measures may involve only facility-based revisions, and some may require corporate engineering or corporate environmental department input. An implementation action plan should be developed, and periodic status updates should be required from responsibility assignees. Requirements for periodic environmental remedial action reports and updates should receive at least the same priority as production and/or financial reports.

WASTE MINIMIZATION

Introduction

This section is titled "Waste Minimization." But in the current regulatory climate, this term is a misnomer because "pollution prevention" is the term currently used by the EPA as they address approaches to reducing wastes and achieving zero discharge goals. By original EPA definitions, waste minimization means the reduction, to the extent feasible, of any solid or hazardous waste that is generated or subsequently treated, stored, or disposed of. Waste minimization techniques focus on source reduction or recycling activities that reduce either the volume or the toxicity of the waste.

Pollution prevention is an extension of the concept of waste minimization. While pollution prevention includes waste minimization, it broadens the concept to include minimizing the generation and release of all hazardous materials and wastes to all environmental media. Many of these issues are addressed in the EPA *Pollution Prevention Benefits Manual,* Volume I. This manual was prepared for the office of Policy, Planning and Evaluation and the Office of Solid Waste of the EPA, October 1989. The EPA created the Pollution Prevention office, which will be the focal point for an integrated, cross-media approval to pollution prevention. In an excerpt from the Federal

Register of January 26, 1989, the EPA notes that "...developing and implementing a new multi-medial prevention strategy, focused primarily on source reduction and secondarily on environmentally sound recycling, offered enormous promise for improvements in human health protection and environmental quality and significant economics benefits."

This new approach is quite different from the EPA's past efforts to protect human health and the environment, in that it recognizes that many of the benefits of managing and controlling waste have already been achieved. Therefore, further environmental gains must come from preventing the generation and release of pollutants.

Many states are also moving towards concepts of pollution prevention. In a document, *Pollution Prevention Initiatives in New Jersey,* prepared by the New Jersey Department of Environmental Protection and Energy, office of Pollution Prevention, October 1991, Pollution Prevention is defined as "...any method or technique which reduces or eliminates the use or generation of hazardous substances by focusing on the front end of a production process." Pollution prevention includes material substitution, product reformulation, production process redesign or modification, in-process recycling, and improved operation and maintenance of equipment. Note that the principal effects of pollution prevention are directed towards source reduction. The emphasis is on preventing the generation of waste in the process itself. The principle and goals of pollution prevention and waste minimization are specifically addressed in the CAA Amendments.

The Clean Air Act Amendments

As discussed earlier in this chapter, the CAA Amendments authorized the EPA to define RACT and MACT for VOC and HAPs reduction, respectively. The EPA can also require raw material substitutions or specific technologies to minimize emissions (waste). SARA and the Waste Minimization and Control Act of 1989 also have waste minimization goals. Thus, it is exceedingly important to follow up all audits with a waste minimization program, if the data indicate that one is needed.

Pollution prevention has several triggers. Disposal site availability or lack thereof, land disposal restrictions, disposal costs, regulatory compliance requirements, public pressure, and/or corporate policy are all operative factors. In addition, environmental noncompliance has severe cost penalties in terms of dollars and civil and criminal liabilities. It is our firm belief that pollution prevention is more cost-effective than any other method for dealing with process emission (wastes) when long-term liability is considered. If a waste is not made, it cannot come back to haunt you when a storage, treatment, or disposal site becomes a Superfund or RCRA corrective action site.

A pollution prevention program requires systems audits, review of results, selection of options, cost evaluation of alternatives, a review of regulatory constraints, and risk assessment to prioritize waste toxicity. The program must be tailored to each site or even process, but the basic components remain the same. Figure 10-2 is a chart of options for pollution prevention. As is shown in the case history listed below, a major component of pollution prevention is changes in work practices. Not all actions require high technology, although significant technology changes are required to obtain maximum results. To be successful, changes in work practice will require significant investment in operator training. One final point, "process," as used in this section, refers to any unit operation in a facility that generates a waste. Chemical manufacturing processes, steam generation, wastewater treatment, cooling tower blow-down, abatement device operation, repair and maintenance operations, and others are all processes that should be considered for waste minimization.

Reduction in waste *generation* is the crux of a successful waste reduction program. This approach differs from RCRA, which addresses solid hazardous waste management *after* the waste is generated. This approach requires significant knowledge of the process as well as in-depth evaluation of potential process changes. Raw material substitution, by-product reuse, catalyst changes, equipment modifications, changes in operating parameters, continuous vs. batch operation, discontinued manufacture of a product with a high pollution profile, and shut down of a process that either is too expensive to fix or cannot be fixed are all options to consider. More than one option or a combination of options may fit a given process; then the decision criteria listed below should be used to select the optimum solution.

Each option must be studied in detail to determine what actions are possible and to select the most cost-effective solution. A ranking of minimization options should be done on a criteria basis and the most cost-effective option chosen. Items to consider are:

- Possibility of waste toxicity and toxicity reduction;
- Possibility of waste volume reduction;
- Cost savings;
- Time to implement;
- Effect on operations and on product quality;
- Cost of equipment changes, product testing, and so on;
- Regulatory constraints, and so on; and
- Corporate policies and goals.

Implementation of the selected process changes should be well documented so that results can be assessed and reductions can be proven. Moni-

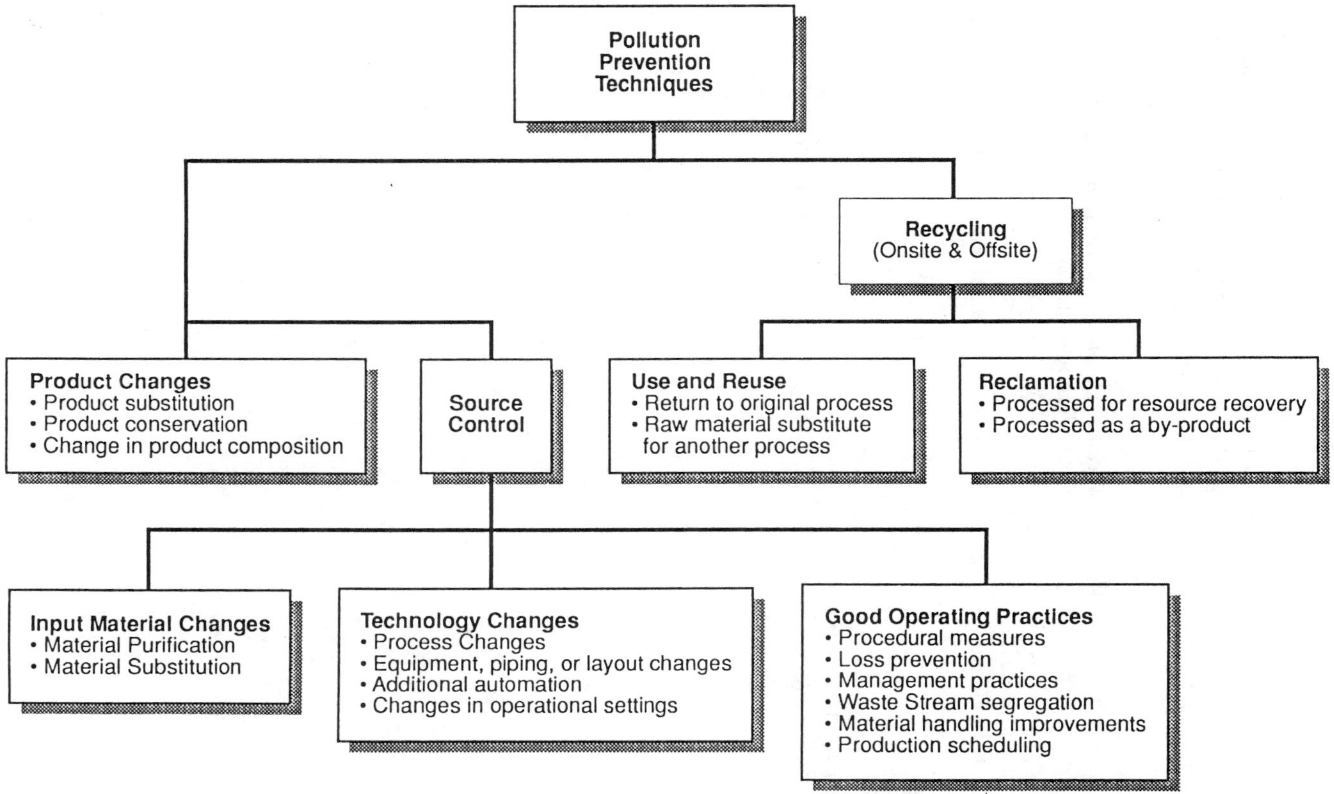

FIGURE 10-2. Pollution prevention options.

toring and evaluation should be done until you are satisfied that the desired results have been achieved. Waste minimization is not a one-time occurrence. Technologies, regulations, costs, disposal site availability, and other factors constantly change. Thus, processes should be periodically reviewed to determine if additional waste minimization can be done.

References

Gimpel, Jean. 1976. *The Medieval Machine: The Industrial Revolution in the Middle Ages,* New York: Holt, Rinehart and Winston, page 82.

BCM Engineers, Inc. 1989. *Fugitive Emissions Study, Alabama Consortium.* Mobile Alabama, January.

Chemical Manufacturers Association. *Pollution Prevention Resource Manual.* Washington, D.C., 1991

Environmental Protection Agency. 1990. *Estimating Peak Releases for the Toxic Release Inventory,* November 2.

Federal Register, Volume 55, Number 61, *40 CFR Part 261 et al., Hazardous Waste Management System; Identification and Listing of Hazardous Wastes; Toxicity Characteristics Revisions; Final Rule,* pages 11795–11877, March 29, 1990.

11

Property Transactions

Jill Barson Gilbert and Douglas J. Hileman

INTRODUCTION

Consider the following scenarios.

A company prepares a Letter of Intent to enter into a $2 billion joint venture involving several refining operations, over 50 terminals, and numerous gasoline service stations. The company has an environmental consultant review the environmental, health, and safety issues associated with the transaction and discovers potential environmental liabilities in the hundreds of millions of dollars. Should the buyer kill the deal?

A light industrial facility built in the 1970s has a leaking underground storage tank. The leak results in soil and groundwater contamination, with the potential for hundreds of thousands of dollars in potential cost. The current site owner is financially incapable of shouldering the costs of remediation. How could an audit or assessment have limited the owner's potential liability?

An airport authority needs more land for a runway expansion. The authority must buy or condemn several properties to proceed with the project. One of the properties has asbestos insulation. What activities must the property owner undertake to minimize the liabilities associated with asbestos during demolition?

Buyers, sellers, and other parties to property transactions may find themselves in these situations if they are not aware of environmental, health, and safety issues. Environmental issues are important in property transactions for several reasons. First, environmental regulations are becoming more stringent and more political. Regulatory agency enforcement policies are becoming more aggressive, holding that ignorance is no excuse for noncompliance with regulations or standard industry practices. The Securities and Exchange Commission is changing how companies must disclose environmental liabilities. There is a continuing search for *deep pockets* to fund environmental improvements and cleanups. Finally, there is growing public support for protecting the environment.

The public demands a clean environment and a safe and toxic-free workplace. A 1989 *New York Times* survey found that over 80% of those polled agreed that "protecting the environment is so important that requirements and standards cannot be too high, and continuing environmental improvements must be made *regardless of cost*" (emphasis added) (Suro 1989). Environmental issues affect not only the major oil and chemical companies, but every company or business handling or working around any quantity of hazardous substances.

Smaller companies may be even more at risk because they seldom have access to internal staff specially trained in environmental, health, and safety regulations and procedures. The regulatory system addressing environmental issues has evolved in piecemeal fashion over the last 20 years and can be complicated and confusing to the non-environmental professional. In addition, if companies building or expanding facilities fail to obtain proper environmental permits, contractors also can fall victim to significant delays.

This chapter discusses key environmental issues, situations, and procedures that may arise during property transactions, and provides information on:

- Types of risk;
- Regulatory framework and dynamics;
- Environmental due diligence;
- The property life cycle; and
- Various environmental issues encountered in property transactions.

ENVIRONMENTAL DUE DILIGENCE

Types of Risk

Environmental risks impact not only chemical plants and refineries, but also such facilities as airports, service stations, restaurants, farms, fertilizer outlets, auto and truck dealerships, and retail stores. Environmental risks may result from past and current activities at a site, and may result from both on-site and off-site operations. The public plays an important role in defining environmental risk. Environmental risk falls into four categories:

1. *Residual risk* is inherent in real property and improvements. One may ask, "If the operations or tenant(s) were gone, what would be left?" Typical

concerns include contaminated soil or groundwater and asbestos in building materials. These risks can affect contractors during the normal course of site work and are the focus of most standard real estate assessments.

2. *Ongoing or operational risk* relates to maintaining operations and includes compliance with rules and regulations that govern environmental areas. Examples of relevant operational concerns are obtaining proper permits for air pollution exhaust and limiting discharges of industrial wastewater into the city sewer. Examples of ongoing risks for a contractor's operations include removing an asbestos-containing roof and storing contaminated soil.
3. *Personal injury risk* involves injury or sickness from exposure to toxic substances, either suddenly (from releases) or slowly (from long-term chronic exposure). These injuries can be sustained on site or by neighbors. For example, a contractor employee may sustain injuries if an underground tank containing solvents is discovered and the chemicals are handled improperly. Another example is a release of acid fumes to a neighborhood near a chemical plant.
4. *Public policy risk* cannot be cited in specific regulations, but often has powerful effects. For example, zoning laws can change and preclude the development of an otherwise encouraged industrial park. Agency officials and elected politicians take stands on certain issues, such as pollution controls or land use, that affect contractor operations.

Public expectations and demands may exacerbate these risks, while Congress and state and local officials create numerous regulations to address key issues. The result is a complicated and confusing tangle of regulations governing environmental (and health and safety) issues. These regulations and the common law interpretations of them through various court cases can affect potential buyers and sellers in many ways. Consequences of violations or excursions from regulatory mandated procedures include significant fines and penalties, restrictions on or denials of projects or land use, work slowdown or shutdown, or, in some instances, criminal prosecution and possible imprisonment.

Congress periodically reauthorizes and amends federal statutes, such as the Clean Air Act and Superfund, often with substantial changes. The Environmental Protection Agency (EPA) or state agencies can make substantial changes or changes of interpretation in implementing regulations. Common law interpretations also influence how environmental issues are handled.

REGULATORY FRAMEWORK

Many federal, state, and local environmental regulatory issues impact the value and operability of a property. Typically, Congress passes each regulation in reaction to a specific need or situation. Regulatory amendments and policy implementations contribute to growing, complex requirements that have corresponding effects on the entire regulated community. The following sections examine several federal environmental statutes, with selected state counterparts, that relate to environmental risks and liabilities linked to property transactions. Rather than presenting a tutorial on one or more regulations, individual sections outline the history and dynamics of regulatory policies and their implications, as well as areas currently requiring attention.

Clean Air Act

The Clean Air Act (CAA) Amendments of 1970, with further amendments in 1977 and 1990, was the first major federal statute of the environmental movement as we know it. The CAA Amendments responded to highly visible billowing smokestacks and smoking auto tailpipes by creating a set of National Ambient Air Quality Standards (NAAQS) for a variety of relatively common *criteria* pollutants. The CAA Amendments included a regulatory framework requiring various levels of air pollution control technology, and they especially targeted industrial sources. Air pollution sources in areas that failed to meet the air quality standards were required to install additional pollution control systems.

The 1977 CAA Amendments also included provisions to identify hazardous air pollutants. The National Emissions Standards for Hazardous Air Pollutants (NESHAPs) set strict and explicit requirements for handling these hazardous air pollutants and preventing exposure of workers and the public. Initial public and Congressional expectations were that scores of chemicals eventually would be subject to this strict regulation. However, several years after the program went into effect, the EPA had designated less than a dozen chemicals as NESHAPs pollutants. Among them are asbestos, commonly used in building materials until 1979; benzene, a constituent of gasoline; and formaldehyde, a component of some building insulation.

As air regulatory programs dealt with the more obvious pollution sources, health risk evolved as a growing concern, driving air pollution regulations. Air toxics now are regulated with increasing stringency. The 1990 CAA Amendments identified 189 hazardous air pollutants, mandated a broad new permitting system, and introduced additional economic incentives for pollution control.

Many severely polluted areas already have invented creative ways of regulating and controlling air pollution. California (particularly Southern California) has been especially thorough and may provide a glimpse of a national regulatory scenario for air pollution control in years to come.

Proposed development projects in the South Coast Air Quality Management District must consider the air impacts not only of the factory or business itself, but of mobile air emission sources, such as traffic. If contaminated soil is uncovered during the construction of a project, it must be covered with impermeable plastic to prevent volatile substances from polluting the air. A great deal of attention is paid to record keeping and documenting all air pollution–causing materials. For example, businesses repairing airplanes or repainting vehicles must keep running inventories of paint used and must determine their air emissions by performing material balances on their operations.

It is fair to say that air pollution still is an overriding concern for the nation's voters, regulators, and politicians. The regulated community can expect air pollution issues to become more stringent and to have wider effects on operations.

Clean Water Act

Congress passed the Clean Water Act in 1972, also in the early days of the current environmental movement, when the nightly news brimmed with images of sudsy, polluted rivers and dead fish. Rumors circulated about the death of Lake Erie. Regulations required polluters discharging directly to the nation's waterways to treat their wastewater streams to prescribed standards. As these *point sources* installed their treatment systems, regulators turned their attention to municipal systems and the problems associated with the industries that discharged to them. Municipalities were required to "lower the boom" on any industries discharging to the system, requiring them to limit their discharges by treatment or process modification, and subjecting them to hefty surcharges or fines for excessive discharge of pollutants.

After a series of notable oil spills into the nation's waterways, legislators amended the Clean Water Act to require taking measures to prevent the release of any type of oil into a waterway. The Act requires a written Spill Prevention, Control and Countermeasure Plan for all facilities that store 1,320 gallons or more of oil aboveground, or more than 660 gallons in a single tank. The regulations broadly define *oil.* The definition includes frequently used materials, such as diesel fuel, dust suppression oil, fuel oil, machine oil, and others.

Stormwater is one focus of current regulatory activity for water pollution. Rainwater that comes in contact with hazardous substances can entrain them, and a subsequent flow off site can cause pollution of waterways, sediments, and groundwater. These hazardous substances may include roofing compounds, oil stains in parking lots, pesticides used in agriculture, or any number of other materials. Facility owners and operators may have to include extensive systems in building projects to handle stormwater and to prevent run-on from neighboring facilities. The EPA has promulgated regulations for stormwater discharge permits, with the possibility of controls in the future.

Another evolving focus is pesticide contamination of surface water sediments or groundwater. Site operators may need to implement special handling procedures if they are involved in projects where land is converted from agricultural to other uses.

Resource Conservation and Recovery Act (RCRA) and Hazardous and Solid Waste Amendments of 1984 (HSWA)

Congress passed the Resource Conservation and Recovery Act (RCRA) in 1976 in the wake of public outrage about Love Canal, with the goal of establishing a system to regulate wastes *from cradle to grave*. The EPA took nearly 4 years to develop regulations to categorize over 400 commercial chemicals as *hazardous* if discarded and to designate over 100 waste streams as hazardous by definition.

Hazardous waste generators must characterize all waste streams for proper disposal. This requires knowledge of the source, composition, and chemical and physical attributes of every waste. Generators must determine if the wastes are RCRA hazardous wastes, either by being listed or by showing certain characteristics. *Listed* wastes are hazardous by definition if they originate from a source described in the regulations. Two examples are spent halogenated (chlorinated) solvents from degreasing operations and leaded gasoline tank bottoms from petroleum refining. *Characteristic* wastes are hazardous if they meet one of four criteria:

1. Ignitability;
2. Corrosivity (pH less than or equal to 2.0 or greater than or equal to 12.5);
3. Reactivity (explosive, reacts violently on contact with water, or can emit toxic gases, such as sulfide or cyanide); and
4. Toxicity (the waste extract exceeds specified concentrations of any 1 of 8 heavy metals, 6 pesticides, or 25 organic compounds).

In 1990, the EPA changed the test defining RCRA toxic wastes from the Extraction Procedure (EP) test to the Toxicity Characteristic Leachate Procedure (TCLP). RCRA toxic wastes now include many waste streams with organic constituents that previously were not regulated. Testing performed on many waste streams before May 1990 may not suffice for current standards. Regardless of whether wastes are determined to be hazardous or nonhazardous, the waste generator must maintain documentation supporting the waste classification.

The RCRA also categorized entities that handle hazardous wastes as generators, transporters, or treatment, storage, and disposal (TSD) facilities. Generators were given responsibilities for EPA notification, classification, ensuring proper management, and record keeping. *Generators* include anyone who arrange for the transportation or disposal of wastes. Generators also retain ultimate permanent responsibility for those wastes, wherever their final disposal may be. Transporters must register with the EPA and meet specific requirements for training, vehicle inspection and maintenance, record keeping, and work practices. TSD facilities are confronted with complex permitting provisions, including extensive written management procedures, descriptions of activities that will be required at partial or complete facility closure, and proof of financial stability to cover all closure costs. The retention of wastes classified as *hazardous* for over 90 days triggers the full impact of TSD permitting requirements. This RCRA provision has great potential to impact those who inadvertently exceed this 90-day limit.

In 1984, the Hazardous and Solid Waste Amendments (HSWA) added a broad scheme to regulate underground storage tanks (USTs). Requirements include phased-in technical standards for USTs, annual leak-testing of tanks and associated piping systems, installation and maintenance of leak detection systems, proof of financial ability to cover closure costs, and prompt reporting requirements and remediation of any leaks. These stiff requirements are a major factor contributing to the decrease in the number of corner gas stations.

Other changes in RCRA focus on waste minimization, land disposal restrictions, regulating smaller quantity generators (currently 220 pounds per month in some states; zero threshold in others), handling empty containers, regulating recyclers, controlling boilers and industrial furnaces that burn regulated hazardous wastes, regulating medical and infectious wastes, and restrictions or bans on hazardous waste exports.

Many parties to property transactions may encounter hazardous wastes in normal operations, such as in facility renovations or demolition, in subsurface contamination, or in other ways. A party unwittingly can become a generator of hazardous waste by arranging for its disposal or by signing a shipping manifest. One needs only to carry hazardous wastes, such as certain contaminated soil, on any public thoroughfare (even if moving it from one portion of a construction site to another), to be considered a hazardous waste transporter. Similarly, one needs only to store hazardous wastes such as used degreasing solvents, waste paints, or qualifying contaminated soil for more than 90 days to be considered a TSD facility. Fines authorized by hazardous waste regulations normally are $500 to $5,000 *per day of violation*. Enforcement agencies can easily prove violation through documentation. It is entirely the generator's responsibility to determine if the nature of wastes being handled fit the defined criteria of hazardous wastes.

Comprehensive Environmental Response, Compensation and Liability Act (CERCLA or "Superfund"), and Superfund Amendments and Reauthorization Act of 1986 (SARA)

Although regulations had "closed the loop" on air, water, and hazardous waste pollution by 1976, the public remained concerned about addressing abandoned hazardous waste facilities. Cost estimates for cleanups varied, but all were in the hundreds of billions of dollars.

Congress passed Superfund in 1980 to address the issue of paying for the cleanup of abandoned hazardous waste sites. Superfund included strict liability provisions, exposing any owner of real property to all the risks thereof, including soil and groundwater contamination, even if the contamination originated off site or was left by former owners or operators. Provisions allowed for cost recovery from any *owner or operator*, which can include any person or entity involved with placing, moving, or releasing hazardous substances on the property.

Superfund also included *joint and several liability* provisions. This means that, if a generator sent hazardous waste to a disposal site that required cleanup at a later date, the generator could be held liable for any portion or all of the overall cleanup at that disposal site. A generator of waste placed in a disposal site that later requires cleanup may be designated a potentially responsible party (PRP) for that site, which could lead to significant litigation costs and contributions to clean up. This liability can be proportionally far greater for generators than their original contribution to the contamination. This provision could affect contractors if they had generated (or agreed to stand in for a project proponent as a generator) contaminated soil that was placed in a landfill that later required cleanup. The *EPA can encourage private parties to perform cleanup, or the Agency may perform the cleanup.* Government-managed cleanups typically cost two to three times those managed by the private sector. According to the EPA, PRPs today are managing over 60% of all cleanups in progress, versus 37% 4 years ago (JAWMA 1991). The EPA believes much of this shift in responsibility is a result of changing EPA enforcement strategies (JAWMA 1991).

Superfund also created requirements for reporting any release of broadly defined or listed *hazardous substances* into the environment to government agencies. This was an effort to prevent the occurrence of additional abandoned waste sites. The regulations specified reporting thresholds for all hazardous substances; for many, spills in excess of 1 pound require reporting. Some states, including Texas, have a zero threshold for many chemical spill reporting requirements if the spill has the potential to impact the environment.

Several states created their own superfunds to augment the federal effort. The latitude of some of these programs goes well beyond Superfund in the ability to order the investigation or clean up of sites.

There are now nearly 1,200 designated Superfund sites, but Superfund has been responsible for the cleanup of disappointingly few of these sites since 1980. The EPA reports that final cleanup was achieved at 63 sites by the end of fiscal year 1991, but expects to have cleanup completed at 200 sites by the end of 1993 and 650 sites by the year 2000 (JAWMA 1992). Superfund trends include:

- Broader circle of PRPs in an effort to pay for cleanups of landfills and regional contamination;
- Protracted, costly, complicated litigation;
- Increased political pressure for making sites cleaner, faster;
- Smaller businesses becoming involved without realizing it through improper management or documentation of small quantities of hazardous wastes;
- Spiraling costs; and
- Continued restrictions imposed by lending institutions or investment or development firms.

State Property Transfer and Disclosure Requirements

New Jersey passed the Environmental Cleanup Responsibility Act (ECRA) in 1983 to expedite the process of cleaning up contaminated sites. Before the close of escrow on an industrial property, the seller must prove to the state's satisfaction that the site is acceptably clean. Regulated properties include Standard Industrial Classification (SIC) Codes 22–39 inclusive, 46–49 inclusive, and some 51 or 76, as designated. Any excessive contamination discovered must be cleaned up, or funds must be provided to the state's satisfaction, before the close of escrow. The state's supervisory role is funded entirely by the parties to the transaction. Several other states, such as Indiana, Illinois, and Connecticut, have laws with similar goals, but none are as rigorous as the New Jersey law.

Massachusetts has the right to impose a lien on a property or a business to recover costs incurred by the state for site remediation. This *Superlien* takes precedence over all other liens, including mechanics' liens that may be imposed by contractors for any type of work on a property.

California is among the states that require disclosure of the impact of hazardous substances on the property, whether it is *known or should have been known*. This requirement applies to property sellers and to tenants vacating the premises.

Toxic Substances Control Act (TSCA)

Congress passed the Toxic Substances Control Act (TSCA) in 1976 to prevent new substances from entering the marketplace without previous testing for potential hazards. Because of political convenience and timeliness, TSCA also included a broad scheme to regulate the use, management, and disposal of polychlorinated biphenyls (PCBs), a group of compounds used extensively since the 1950s in electrical insulating fluids for transformers, capacitors, and other applications.

The need to regulate asbestos was addressed by passing the Asbestos Hazard Emergency Response Act (AHERA) as Title II of TSCA in conjunction with the TSCA reauthorization in 1986. AHERA prescribed testing for asbestos in schools and provided a framework for training requirements for persons engaged in sampling, managing, maintaining, or abating asbestos or any asbestos-containing materials (ACM).

Title III was added to TSCA in 1988 to address radon. TSCA also directs the EPA to develop model construction standards and techniques for controlling radon levels in new buildings. TSCA requirements affect contractors every time building equipment or materials are encountered that *may* contain PCBs or asbestos.

The EPA currently is taking a more proactive and comprehensive role in environmental management by encouraging pollution prevention, rather than by focusing exclusively on controlling pollution that already has been generated or cleaning up orphaned dump sites. In early 1991, the EPA indicated its intention to use the broad authority granted to it under TSCA and SARA (see Superfund discussion) to enforce its policies.

Federal Insecticide, Fungicide and Rodenticide Act (FIFRA)

Congress passed FIFRA in 1947 and amended it in 1988 to regulate virtually all uses of insecticides, fungicides, and rodenticides (pesticides), including:

- Product labeling;
- Product uses;
- Methods of application;
- Licensing of persons applying or managing these substances;

- Worker protection during application; and
- Holding times after application and before reentry into areas.

FIFRA provisions affect several activities that may occur on active or vacant properties. These include spraying weed control products as a service (even on a small scale), authorizing employees to enter a building after it has been sprayed for termites or insects, and engaging in rodent or vector control as a service to shopping center owners.

RCRA and/or Superfund regulate the management of pesticide-contaminated soils, sediment, or residue that one may encounter when developing of agricultural land.

ENVIRONMENTAL DUE DILIGENCE

Due Diligence Objectives

As a property, building, or business goes through its life cycle, environmental risks change because of modifications in operations, regulatory requirements, public expectations, and the involvement of various parties. Property transactions often involve transferring environmental risks to a new party. Environmental risks are of particular interest in the following transactions:

- Purchase or sale of a property, with new principals and financial backers;
- Purchase or sale of a business, with new principals and financial backers;
- Creation of a new subsidiary or joint venture with additional capital from a new investor;
- Involvement of a trust, whether through creation, change in management, or disbursement;
- Purchase of insurance policies to cover losses from environmental damages;
- Refinancing of existing operations/properties;
- Transfer of financial packages to the secondary market; and
- Acquisition of property by eminent domain.

Environmental regulations impact numerous types of property transactions, regardless of whether the interested party is the buyer, seller, or financial institution. Transactions such as mergers and acquisitions, divestitures, buyouts, and refinancing all can benefit from environmental *due diligence*. Environmental due diligence can help those involved in such transactions to identify and manage risks, limit liabilities, and enable the business community to make more informed decisions. The objectives of due diligence are:

- Identify potential environmental risks;
- Quantify potential financial liabilities;
- Estimate the probability of occurrence; and
- Estimate the timing of expenditures.

Phase 1, Phase 2, and Phase 3 Environmental Assessments

Environmental due diligence assessments must address diverse considerations, such as previous site use, proximity to landfills and schoolyards, soil chemistry, and many other factors. This differs from a comparatively straightforward civil engineering analysis of soil load-bearing capability or from an economic analysis of benefits presented by freeway access. To obtain reasonable information with limited effort, environmental investigations are performed in phases. Environmental reviews typically include up to three phases in the field. Sometimes the review includes a fourth phase, designated Phase 0, to collect information available in the public domain before on-site activities. This section discusses each phase in more detail below; each can be summarized neatly in one or two words:

- Phase 0—Remote Sensing
- Phase 1—Risk Assessment
- Phase 2—Field Investigation
- Phase 3—Remediation

Phase 0—Remote Sensing

Phase 0 remote sensing typically is conducted confidentially, without contacting the owner or operator of the subject site. Phase 0 may provide useful historical information and, in a global sense, identify potential problems associated with a property. This phase serves as an initial screening activity to focus the Phase 1 and Phase 2 efforts that follow. Those conducting Phase 0 assessments review publicly available information, such as aerial photographs, property chain-of-title, environmental agency records, and other published data, such as lists of neighboring water wells and local Superfund sites. If there is no Phase 0, the assessor may refer to some of these data sources during the Phase 1 assessment.

Phase 1—Risk Assessment

Phase 1 risk assessments are requested as a fairly standard practice for most transactions of industrial and commercial real properties. Phase 1 projects assess several factors to determine the residual and operational risk of a property:

- Current and previous operations;
- Chemicals used on site and their management practices;
- Regulatory or enforcement activities; and
- Possible effects of neighboring sites.

Those conducting Phase 1 projects may use a combination of data sources, including interviews with site management and operations staff (where available), site inspection, and records review. The Phase 1 assessment findings are used to determine the relative degree of risk and to determine whether further activities, such as a Phase 2 investigation, are appropriate (see Figure 11-1).

The Phase 1 assessor may obtain information by compiling research from a variety of sources, including agency lists, building records, historical aerial photographs, and environmental permits (see Table 11-1). The Phase 1 risk assessment also might include a legal review to determine the extent of protection provided by indemnifications, warranties, and representations or other language in sales agreements.

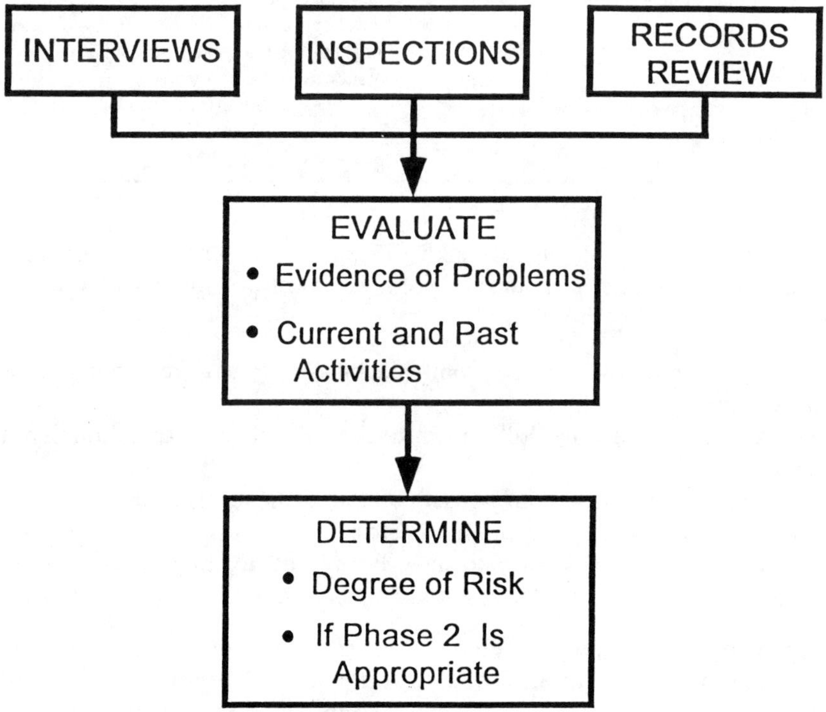

FIGURE 11-1. Phase 1 risk assesment.

TABLE 11-1 Phase 1 Data

Factor	Data Sources
Current and previous operations	Interviews Site inspection Aerial photographs Property chain-of-title Agency records of previous oil and gas exploration activities Process flow diagrams, plans, and specifications
Chemical use and management practices	Interviews Site inspection Facility records
Regulatory or enforcement activities	Permits and registrations Facility records Correspondence with agencies Agency files—compliance and enforcement Interviews
Effects of neighboring sites	Site inspection Aerial photographs Local hydrogeological profile Agency lists—nearby wells Agency lists—Superfund sites Agency lists—underground storage tanks

An experienced professional can evaluate these factors to highlight areas of known or potential problems or risks. Areas of possible concern could include the following situations:

- Underground tanks have been removed, but the status of remaining soil is uncertain.
- The facility is near a landfill known to have problems with contaminant migration.
- Previous tenants left various unused, outdated chemicals on site.
- Structural materials may contain asbestos.
- A previous tenant is known to have been cited for improperly storing hazardous materials, but the residual impact to the property is unknown.

The persons requesting the Phase 1 assessment and those conducting the assessment must agree to the scope of work and the level of significance of the findings. For example, on a $500 million transaction, environmental issues that cost $10,000 to fix may be considered insignificant. However, on a

$50,000 transaction, the same $10,000 issues could substantially affect the property value and/or the future operations, possibly "killing the deal."

Phase 2—Field Investigation

Phase 2 field investigations involve physical sampling and testing of some kind and generally are undertaken to address known or potential problem areas identified in Phase 1. Phase 2 investigations may be initiated if required by law or regulation, agreed upon in terms of a contract, or at any time by an owner and/or operator seeking additional information. The components of Phase 2 include evaluating available information, planning and conducting the field investigation, and determining the extent of contamination and estimating cleanup costs (see Figure 11-2).

The evaluation of existing data is critical to planning the Phase 2 investigation. If extensive soil and groundwater data are available for a site, the Phase 2 activities might be designed to supplement these data. If little or no soil and groundwater data exist, the Phase 2 work plan could be very different.

Specific media that can be physically tested include soil (to any depth), soil vapor, groundwater, surface water, surface water sediment, drinking water supply, indoor air, ambient air, dust, building materials, and residues. Analytical parameters are selected to address the item(s) of concern. Sam-

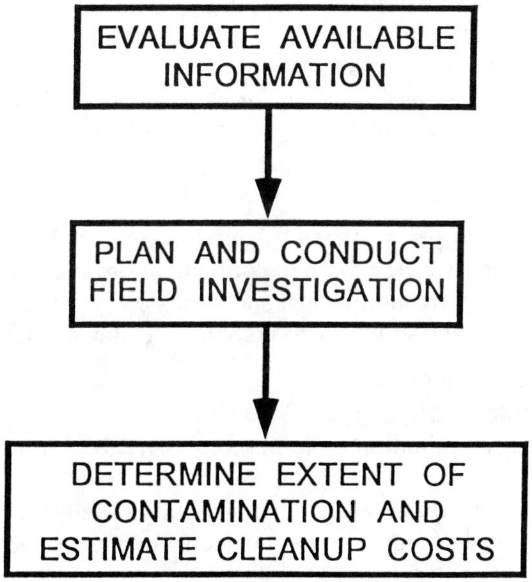

FIGURE 11-2. Phase 2 field investigation.

pling procedures must be followed and documented carefully; physical samples, once taken, have a specified shelf life or holding time and, if not analyzed promptly, may become worthless.

Perhaps the most important part of any Phase 2 investigation is determining a project's goals, scope, schedule, and budget. All four items must be compatible for a successful project. Constraints on any one of these can significantly alter the project outcome. For example, if the Phase 2 budget is $100,000, many soil samples can be collected and analyzed for numerous parameters. If the Phase 2 budget for the same site is restricted to $20,000, perhaps only selected soil samples can be analyzed and conclusions may be less certain. If the constraint is a restriction of access to the site, fewer soil borings may be possible; this reduces the scope and budget and, again, narrows the assurance provided by any conclusions. Or, if the project must be completed in 21 days, cost surcharges may be incurred for drillers, laboratories, and consultants.

Field activities involve many uncertainties and can present challenges or impossibilities that affect costs or the overall viability of a project. Problems could include any of the following:

- Presence of underground utilities, complicating proper placement of soil borings;
- Limited access because of structures or overhead wires;
- Permitting requirements for field activities;
- Agency inspections or requirements during field activities;
- Improper or incomplete documentation on required OSHA training for hazardous waste workers; and
- Discovery of contaminants, odors, or physical conditions not anticipated before commencement of field activities.

Phase 3—Remediation

Phase 3 remediation projects are instigated to clean up environmental contamination, to remove the source of contamination, such as a leaking tank, or to install prevention, monitoring, and detection systems. Phase 3 projects are customized to each situation and require specific information on:

- Type(s) of contamination;
- Extent of contamination;
- Location of contamination (whether impact has been on soil, groundwater, etc.);
- Satisfactory levels for cleanup, whether established by regulations, informal discussions with agencies, on advice from counsel or consultants, or agreed upon in contract; and
- Appropriate technologies.

Acceptable cleanup levels should be well understood before any remediation activity begins. These levels may be established by regulations, policy, informal discussions with agencies, on advice from counsel or consultants, or agreed upon in contract.

Selecting appropriate remediation technology becomes especially challenging when more than one contaminant is present and treatment for the contaminants requires more than one treatment method; access to subsurface contamination is limited by structures, equipment, or traffic patterns on site; contamination has migrated off site; or contamination originated from off site.

Preventive measures may be selected when hazardous materials do not present an imminent hazard. An example of preventive measures is a vapor barrier or passive vapor venting system for a building located in a risk area for subsurface accumulation of methane.

Those involved in property transactions should be aware that there are many companies and consulting firms that provide these services. An experienced practitioner or consultant can focus very quickly on risks associated with technical, management, or potential third-party claims issues. An experienced professional also will design an overall program in phases so that only information specific to relevant risks is reviewed. This minimizes the risk of involvement in irrelevant contamination issues and unnecessary research projects. Practitioners and consultants vary considerably in their experience, management philosophy, and technical capabilities. Those procuring environmental services or working with environmental consultants should invest the time and effort in an in-depth interview to establish confidence that the services rendered will be proficient, equitable, and appropriate to the need.

Overcoming Potential "Deal Killers"

The quality of Phase 0, Phase 1, Phase 2, and Phase 3 Assessments depends upon the time available to complete the assessments, the quality and quantity of existing data available, the cooperation of the seller (or owner/operator of the site), the budget, and other factors. Those hiring professionals to conduct these assessments, whether buyers, sellers, financial institutions, attorneys, or other parties, should be aware that assessments are not guarantees. Assessments are designed to allow these parties to make better business decisions and to structure the transaction agreements appropriately.

Well-designed environmental due diligence can assist the parties to a property transaction in overcoming potential "deal killers." Environmental risks and liabilities may be the *key* issues in chemical plant and refinery transactions. They may be the major concern in other transactions, as well. Buyers, sellers, attorneys, and financial institutions, armed with environmental due diligence results, can prevail over potential "deal killers" by:

- Negotiating indemnification language for past and future environmental liabilities;
- Agreeing to a method of sharing those environmental liabilities that fall into "gray" areas;
- Negotiating the price of the property to reflect environmental contamination or other concerns;
- Setting aside reserves to fund environmental activities, such as extensive investigation and cleanup or participating as a Superfund PRP; and
- Excluding certain sites or tracts from the transaction due to environmental concerns.

Lender Liability

Transactions involve the transfer of environmental risk along with the assets in question. All parties should be interested in accurately assessing the risks and in having a reasonable and clearly defined mechanism for apportioning risks.

- *Buyers* want to avoid acquiring liabilities associated with contamination.
- *Sellers* want to document preexisting conditions to minimize their exposure after the sale.
- *Brokers* want to keep their clients satisfied and avoid issues of nondisclosure.
- *Environmental agencies* want opportunities to prescribe a cleaner environment.
- *Municipalities and redevelopment agencies* want to avoid being involved in hazardous or environmentally risky sites.
- *Developers* want to avoid excessive site design and permitting costs if environmental considerations make projects unfeasible.
- *Investors and financial institutions* want to determine the exposure of their own assets to environmental risks.
- *Contractors* want to avoid liabilities created by unknowingly generating or managing hazardous wastes or by excessive workplace exposure to employees.

Although Superfund provides secured creditors exemptions from environmental liabilities, in the past the exact conditions for this exemption were not clear. Lenders have been held liable for environmental costs through a variety of theories, including:

- Actual holding of title so that an investment constituted ownership;
- Taking ownership through foreclosure;

- Becoming sufficiently involved in site activities through loan policing or trust management as to be considered an operator; and
- Merely having the option to exercise the right to perform environmental inspections or make environmental recommendations.

As a result, investors and financial institutions have taken the lead in determining the course of environmental investigations. There have been a number of proposals from the EPA and Congress to more succinctly define the secured lender exemption. Recent EPA Superfund rules on lender liability have provided some relief. The rules provide a security-interest exemption as long as the lender does not participate in day-to-day management and environmental decisions.

There is a heightened awareness of environmental issues on the part of financial institutions. Lender involvement may mean having to initiate several types of project requirements:

- A Phase 1 environmental assessment report;
- A *new* Phase 1 environmental assessment report if the existing report is outdated;
- A disclosure of any and all documentation or records regarding environmental issues and the condition of the secured interest (property or otherwise);
- Physical sampling and testing to determine the presence of asbestos;
- Other physical sampling and testing programs;
- Requirement to set aside sufficient escrow funds to address contamination remediation, with appropriate contingencies;
- Periodic inspections of site operations during the life of the loan to determine risks of environmental impact; and
- Environmental agency documentation, where available, of acceptable site conditions or remediation activities.

Lender liability is dynamic with regard to applicable regulations, agency policy statements, and evolving case law. Lender policies on environmental issues vary considerably among institutions and may be frequently modified to reflect new considerations. If the involvement of a lending institution is anticipated on any project, it is prudent to contact all possible lending sources ahead of time to obtain policies or requirements on environmental issues so that no critical issues are overlooked.

Properly translating technical information into terms of risk, certainty, anticipated expenditure timing, and other terms of concern to lenders is an important factor in smooth lender review of projects. Parties to transactions should be aware that there are outside resources who specialize in lender management and consulting and can provide benefits to project management, cost tracking and allocations, and expedited project scheduling.

Innocent Landowner

Superfund created sweeping liability that included company owners and operators, real estate investors, bankers, contractors, municipalities, and other entities. As a consequence, small businesses, such as one selling luggage out of a former gas station, were among those who found themselves liable for contamination. In 1986, the Superfund Amendments (known as SARA) attempted to correct certain inequities by creating an *innocent landowner* clause, whereby purchasers of real property could be excused from financial responsibility for cleanups, provided that they had been duly diligent in performing investigations before acquisition. A buyer contemplating the purchase of a former gas station would be expected to assess the possible presence of fuels, oil, or other similar contamination. If the purchaser were duly diligent in investigating and if metal plating wastes that had no reason to be suspected were uncovered during site grading, the new owner may be able to defer cleanup costs to Superfund or other government funding sources. The innocent landowner clause has not been tested in court. Also, some relief is provided to parties who inherit contaminated property. The standards for due diligence are not contained in regulatory guidelines and would be expected to vary with previous property use.

THE PROPERTY LIFE CYCLE

The property life cycle begins when a developer improves raw or previously agricultural land. As the property matures, the land use and physical improvements often change. The life cycle ends with abandonment, demolition, and site disposition; it then may begin anew.

Unimproved, active, inactive, closed, and divested sites all can represent liabilities. Post operating practices, although commonly accepted and legal at the time, can impact the site or surrounding environment and may eventually require remediation. Current landowners or operators can be held liable for remediation if the past owners are not financially able or if clear allocation of environmental liability was not established at the time of transfer. Determining the potential risks, effectively managing these risks, and allocating the appropriate resources can make the difference between profitable enterprises and unprofitable enterprises with major negative financial impacts.

Several activities that often have hazardous waste and environmental impacts occur at various stages of the property life cycle:

- Developing raw land;
- Expanding or modifying existing facility; and
- Demolishing facility or razing site for disposition.

Hazardous waste and environmental issues differ, depending on the types of activities and the maturity of the property or its improvements. These issues impact land use, residual waste management, building design and operation, and the public image of both the contractor and the property user. Each stage of the property life cycle is discussed below.

Developing Raw Land

Environmental Considerations

Even undeveloped land can present liabilities that can hinder or halt site development if not properly managed. As at any transaction stage, a project proponent should conduct environmental due diligence studies to identify pending or latent liabilities. These studies are critical for the owner, operator, purchaser, or other proponent to identify potential "deal killers" or major expenditures. The level of detail in the due diligence review varies with the types of proposed operations and the sensitivity of the surrounding area. Environmental considerations of special interest on raw or previously agricultural land include:

- Hidden contamination originating off site;
- Pesticide residues;
- Radon;
- Wetlands;
- Sensitive or endangered species;
- Protected surface water or groundwater sources; and
- Unauthorized dumping.

A Phase 2 site investigation may be undertaken to establish baseline physical property conditions or to further delineate any contamination or other uncertainties identified in Phase 1. The Phase 2 investigation for raw or agricultural land typically includes soil or groundwater sampling and analysis. Other Phase 2 site investigation activities that may be appropriate for undeveloped property are a site survey, contact with the local utility company, or a geophysical survey for underground utilities, an inventory of any wastes present from unauthorized dumping, a special site inspection for sensitive or endangered species, and an archaeological or other cultural resources inspection.

Phase 3 remedial actions are less common on unimproved property than on developed property. Nonetheless, remedial actions or changes to proposed projects may be required as a result of Phase 2 findings. These corrections might include:

- Removal of pesticide-contaminated stream sediment;
- Removal of drilling muds;
- Proper abandonment of wells; and
- Project redesign to accommodate site-specific or regional-sensitive species or archaeological remains.

Project Implementation

Project proponents should be aware that the National Environmental Policy Act (NEPA) and state counterparts frequently apply to proposed projects on unimproved property. NEPA provisions may require Environmental Impact Reports (EIRs) to evaluate the environmental impacts in very broad terms. This includes possible effects on flora, fauna, traffic, noise, appearance, and overall community standards of living. EIRs have long lead times and can present considerable delays while the project undergoes review by regional development agencies, city councils, or citizens groups. Generally, EIRs are performed by companies or professionals with differing areas of expertise than those discussed throughout this chapter.

If the environmental due diligence effort does not identify "deal killers" and the project is determined to be worth pursuing, the next step is for the project proponent to identify and apply for any needed environmental permits and registrations for the project itself. Permits often are a critical path item and require lead time before construction or other site activities begin. Federal, state, and local agencies all can be involved in the permitting process, depending on the site location and the type of project proposed. Some activities requiring permits or registrations are air emissions, wastewater discharge (including stormwater runoff), discharge to sewers, solid or hazardous waste management, or the use of aboveground and underground storage tanks (USTs). As one might expect, chemical plants may require years to complete the permitting process. However, distribution centers, office buildings, and other relatively benign facilities may require one or more environmental permits.

Regulations are especially strict, and certain types of facilities require obtaining air permits before construction can begin. The failure to obtain proper air permits for all equipment that eventually will be regulated may bring construction to a standstill, even in very early phases of site grading. The air emissions permit process is more complex in areas that fail to meet the NAAQS (non-attainment areas). The key pollutants of concern in these non-attainment areas are volatile organic compounds (VOCs), particulates, sulfur dioxide, carbon monoxide, and nitrogen oxides. Also, the permitting of toxic air contaminant emissions is more stringent today than in the past.

Wastewater permits also may be a project prerequisite. If process wastewater or stormwater to be generated by the project will be discharged to a sensitive (or already over-polluted) waterway or to a municipal treatment plant already strained to capacity, the permit may be difficult to obtain. Recent stormwater regulations may require a discharge permit for new construction, as well as for existing facilities that in the past did not need permits.

Other permit applications and registrations should be applied for before construction, but usually do not pose an imminent threat to the shutdown of construction activity. Underground storage tank registrations and RCRA hazardous waste permits must be approved and received before start-up of operations. If permits are not in hand early, however, agencies could change requirements, thus affecting project design and construction and causing work slowdowns.

Public opinion frequently influences the successful outcome of environmental permit applications. If the site operations are controversial or can affect the community, project proponents should begin working with community leaders early in the planning process to determine specific public concerns. Generally, close contact with the community throughout the planning, permitting, and construction processes can make the project run more smoothly.

Before beginning field work, contractors should review proposed site activities and develop environmental contingency plans to address how to manage materials handled during construction or remediation. Plans should cover contingencies for spills and physical and health risks from toxic or hazardous materials or wastes. If environmental permits are required for the construction activity, environmental contingency plans may have to be in written form. Regulations in some jurisdictions simply specify a general performance standard prohibiting pollution.

Expanding or Modifying an Existing Facility

Environmental Considerations

Expanding or modifying an existing facility can present liabilities because of the impact from preexisting operations or the nature of proposed activities. Plant changes involving modifications to existing equipment, for example, can change the nature or amount of emissions to the environment or may pose new risks to workers during construction or after modification. Physical plant expansion can impact the environment and the local population with effects that include air emissions, stormwater runoff, emergency evacuation routes, and noise.

Project proponents generally conduct environmental due diligence assessments before facility expansion or modification to minimize the environmental impacts and liabilities. Environmental due diligence studies at this stage of the property life cycle can indicate whether past practices have resulted in contamination or whether special precautions must be taken in project design or construction as a result of preexisting conditions. Environmental due diligence for expansions or modifications may include one or more phases as described in the section on Due Diligence.

Phase 1 site assessments that are performed for site modification or expansion may emphasize:

- A review of the potential environmental impact to soil or groundwater from existing operations;
- A review of the potential environmental impact to soil or groundwater from any known previous operations;
- Changes in land use that may trigger different environmental standards;
- Changes in regulatory standards for cleanup levels that may resurrect previously "completed" site remediation;
- Asbestos that may be disturbed during construction;
- Previously closed sites that may be disturbed during construction; and
- Discussions with federal, state, and local environmental agency officials to review the overall compliance history of site operations.

Phase 2 site investigations become more important at the expansion/modification stage of the property life cycle, serving as baseline studies to document conditions at the time of expansion, reducing the possibility of future enforcement problems. This benchmarking is especially important if a new owner or operator is entering the picture. It is also critical when the proposed project will continue to use hazardous materials on the site, creating the possibility of commingling liability for soil or groundwater contamination. Phase 2 investigations that might be especially appropriate at the expansion/modification stage of the property life cycle could include:

- Soil gas survey;
- Soil samples from shallow subsurface or deeper borings;
- Groundwater samples;
- Sampling of building materials for asbestos; and
- Geophysical surveys for underground utility lines.

Phase 3 site remediation projects may be required before or in conjunction with site expansion or modification. If the presence of residual materials or site contamination is identified in Phase 2, the proper allocation of responsibilities, risk, and liabilities is critical. Selecting the appropriate remediation technologies is a factor in implementing Phase 3 projects and must consider aspects of the project that include:

- The chemical nature of the contaminant;
- The lateral and vertical extent of contamination;
- Whether soil or groundwater has been impacted;
- Whether site process operations will continue throughout construction activities; and
- The ease of obtaining permits from relevant environmental agencies.

Some remediation technologies, such as the excavation of soil contaminated with heavy metals, may be disruptive to existing site operations and construction activities. Other technologies, such as vapor extraction of dry cleaning solvents from soil, may be installed and operated with very little disruption of continuing operations.

Project Implementation

Changes to an existing facility can alter the nature or amount of air emissions, wastewater discharges, and solid or hazardous waste generation. If the environmental due diligence does not identify "project killers," the project proponent must determine if permits are needed and apply for them. As when developing raw land, permits often are a critical path item and require considerable lead time before construction or modification can begin.

Regulations for air permit modifications are especially explicit in requiring the completion of all aspects of permit applications and approval *before* starting building or structure modifications, even at existing facilities. The expansion of an existing facility may trigger more stringent controls for the entire plant. In air non-attainment areas, the facility may be required to shut down some older emissions sources to offset the emissions from the new or modified equipment. As in the case of developing raw land, permits for changes in project wastewater discharges also may be a critical path item in locations where receiving streams or municipal treatment plants have limited capacities for additional pollutant loading from wastewater.

If hazardous substances will be disturbed during site modification activities, regulatory agencies may require the submittal of written plans detailing the exact procedures and plans for these activities. They may elect to send agency inspectors to observe these activities and verify compliance with the written plan. For example, local agency representatives may collect soil samples during a contaminated soil cleanup to verify that all designated contaminated soils are being removed. Agency inspectors may wish to observe the removal of asbestos-containing materials to verify the appropriate procedures for preventing fibers from becoming airborne and escaping into the ambient atmosphere. Failure to submit proper notifications and plans may expose the project proponent or contractor to fines or to an excessive burden of proof verifying that proper procedures were followed during the course of the project.

As in developing raw land, those working on the site must be aware of and plan for environmental contingencies, physical hazards, and environmental and personnel exposures. If sites are to be expanded or modified while existing facilities continue to operate, contractors must pay special attention to the compatibility of their own operations with ongoing site activities. Ongoing process activities may present many situations that demand extra precautions.

Environmental, health, and safety emergencies can occur during expansions and modifications, so contractors should develop appropriate emergency, contingency, and spill plans. Adequate preparation for emergencies can minimize environmental liabilities.

Demolishing, Razing and Site Disposition

Environmental Considerations

At the end of the property life cycle, the contractor often is involved in demolition or razing to prepare for site disposition. Environmental due diligence once again is recommended at this stage. Ideally, due diligence is performed while the facility is still in operation. Phase 1 and Phase 2 activities depend heavily on the first-hand observations by an environmental specialist of process operations, materials management, and environmental management systems. The availability of facility personnel who are familiar with current and previous operations and the identity of chemicals used contribute significantly to the proper due diligence investigation. All too often, however, environmental investigations are not initiated until after operations have ceased. Environmental investigators are left to guess at important information, such as the location and types of equipment, chemicals, and procedures used for the previous removal of underground storage tanks.

Demolition or razing activities have many of the same concerns as those for new construction, expansion, or modification. In addition, these activities can expose toxic or hazardous wastes or materials that otherwise might not be a concern. Some materials, such as asbestos, do not meet today's construction and environmental standards and have special handling and disposal requirements. Project proponents should know if these materials are present before starting demolition and conduct any required inspections, sampling, and testing of the building components or wastes. Prudent management reduces liability, positively affects the property value, and reduces waste management costs.

Phase 1 assessments at the demolition/disposition phase of the property life cycle include special consideration of the following issues:

- Presence of asbestos-containing materials in portions of the building to be razed;
- Documentation of previous site investigations or tank removals;
- Presence of underground storage tanks, sumps, oil/water separators, clarifiers, or floor drains that may trigger regulatory requirements for abandonment; and
- Facility plans for discontinuing operations and disposing of leftover residual materials and wastes.

Phase 2 investigations at this stage of the property life cycle would be similar to those for an expansion/modification project. The testing of building materials for asbestos and of electrical equipment for PCBs would be expected to be particularly aggressive. Residual materials, tank bottoms, and other residual wastes also would require testing (see next section). Some residuals, if left in process equipment, tanks, or storage containers, become hazardous wastes after a specified period and require special handling.

Phase 3 remediation programs would be similar to those for an expansion or modification. The proper allocation of responsibilities, risks, and liabilities perhaps is more critical for facilities that have ceased operations than for facilities that continue to operate. For example, removal or remediation of residual wastes can be a major concern at this time, if there is no owner or operator to fund such activities as extensive soil and groundwater remediation or costly asbestos removal.

ENVIRONMENTAL ISSUES AND PROPERTY TRANSACTIONS

Those involved in property transactions may encounter many environmental issues during their involvement on various projects, whether they are the buyer, seller, or have other interests. Common environmental issues include:

- Underground storage tanks (USTs);
- Active or inactive disposal waste sites;
- Soil and groundwater contamination;
- Asbestos-containing materials;
- Polychlorinated biphenyls (PCBs); and
- Residual materials and wastes.

Table 11-2 summarizes the types of environmental risks associated with these issues, the timing of these issues during the property life cycle, and when they may require assessment or management. The remainder of this section discusses each issue in more detail.

Underground Storage Tanks (USTs)

Many formerly operating sites have USTs that are removed as site operations change or cease. Other USTs previously unknown to any site operator may be discovered during site activities or closure. Leaks from USTs or associated piping systems or overflow from UST filling can contaminate the soils or groundwater, with potential off-site impacts. Contractors removing USTs should be aware that regulations require advance notification. This affects the lead time before UST closure can begin. If the USTs have stored chemicals, soil borings or other assessment activities may be required before tank removal. Removal of flammable vapors is required by tank de-gassing, often with dry ice. Regulations also require UST registration, agency notifications, reporting of contamination associated with USTs, spill response

TABLE 11-2 Environmental Issues Typically Encountered in Property Transactions

		Stage of Property Life Cycle		
Environmental Issue	Types of Risk	Unimproved Land	Expansion/ Modification	Demolition/ Disposition
Underground storage tanks	Residual Ongoing	No	Yes	Yes—Important
Active or inactive waste sites	Residual Ongoing Personal injury Public policy	Yes—Unauthorized dumping	Yes	Yes
Soil or groundwater contamination	Residual Ongoing	Possible	Yes	Yes
Asbestos-containing materials	Residual Personal injury	No	Yes	Yes—Important
PCBs	Residual Ongoing	No	Possible	Yes
Residual materials and wastes	Residual Personal injury Public policy	No	Yes	Yes—Important

procedures, investigation activities, verification of site conditions, and notification upon closure.

Federal and state laws and regulations may require special registrations to legally remove a UST. Special employee training may be required, and documentation of this training may be required to be available on site.

If involved in the excavation of a UST, contractors should check the integrity of the UST and piping after removal for holes, corrosion, or evidence of leaks or overfills. Contractors also should check the excavation for visual or olfactory signs of leaks, and may use a portable organic vapor analyzer to screen for contamination. Even if not required by the agencies, contractors should collect soil samples for analysis to ensure that the excavation is clean before backfilling and restoring the original contours. Removed tanks should be cut apart or perforated to prevent subsequent reuse. A receipt verifying proper disposal should be maintained to minimize future liabilities.

Other structures or facility features may be regulated as USTs in some jurisdictions and may include in-ground clarifiers, in-ground oil/water separators, piping systems connected to USTs, and in-ground trenches or floor drains. These units may be encountered as part of industrial wastewater collection systems, near discharges to city sewers, or as parts of pits used for auto maintenance or waste oil collection. Requirements for investigation, treatment, and remediation may apply in these situations. These features most typically are encountered at facility demolition. For example, if the land use is being changed from commercial auto maintenance to residential, then requirements for investigation and remediation may be even more stringent.

Active or Inactive Waste Sites

During site demolition or razing, those who work in or around active or inactive waste disposal sites should be aware of potential hazards and special requirements. Discontinued operations can include hazardous waste or other solid waste landfills, surface impoundments (ponds), waste treatment units, or drum or tank storage areas.

Hazardous and solid waste management units usually are associated with industrial activities. Waste treatment units include incinerators and biological or chemical treatment systems. Some common waste storage areas are drum storage, roll-off container storage, tank farms, and surface impoundments. Surface impoundments commonly are used to manage cooling water, stormwater, wastewater, and mining overburden. Waste disposal activities include landfarming for biological degradation and landfills. Landfills may be used for wastes ranging from innocuous trash and debris to hazardous wastes; at older industrial facilities, hazardous materials may have been disposed of according to practices of earlier days.

Site owners or operators may hire outside professionals to develop closure protocol or to close hazardous or solid waste management facilities per predetermined plans and specifications. RCRA closure plans address closure of RCRA-regulated areas, including the removal of hazardous waste inventory, equipment decontamination, and site remediation. Closure plans are part of RCRA permits at TSD facilities, but should be thoroughly reviewed as actual closure activities are being contemplated.

Clean closures involve the removal of all wastes and residual contamination to background levels or to cleanup levels acceptable to federal, state, or local agencies. Other closures involve leaving the wastes in place. For example, a clay cap and grassy cover may be added to a landfill. Surface impoundment closure often involves dewatering sludges and stabilizing residues in place rather than excavating all contaminated materials for off-site disposal. Closure with hazardous wastes remaining may require a note in the property deed record that hazardous waste activities occurred, whereas *clean closure* generally will not require deed recordation.

Rigorous requirements apply if field activities involve hazardous wastes. In addition, the requirements vary from state to state and are growing more stringent over time. Before beginning work on any project involving wastes, contractors should be satisfied that the site owner or operator has performed the appropriate tests or evaluations and that it is safe to proceed.

Hazardous and solid waste regulations can impact scheduled field activities and can be a critical path item. RCRA closures must follow a regulatory-driven timetable, including for clean closure the time allowed for waste removal and for site decontamination and final closure. The regulations also dictate procedures for agency notifications. Non-RCRA closures may require similar timetables, depending on the agencies involved.

At major milestones in clean closures, the site owner or operator should conduct sampling and analysis to verify the level of decontamination achieved, regardless of whether the facility managed hazardous wastes during its lifetime. Project proponents may wish to review the sampling and laboratory data or collect samples independently, to minimize future liability from unknowingly disturbing or moving materials and aggravating the problem.

Soil and Groundwater Contamination

During site demolition or razing, there may be signs of soil and/or groundwater contamination from past operations, disposal practices, or substances entrained in stormwater runoff. These factors can contribute to both on-site and off-site impacts. Project proponents should look out for soil staining or discoloration, distressed vegetation, and odors.

Activities such as cleaning process equipment, USTs, or sump removals, if improperly performed, could cause soil or groundwater contamination. If a release occurs, several federal regulations require immediate spill and incident reporting for spills or releases that have the potential to impact human health or the environment.

During site demolition or razing, contaminated soil and debris may be removed for off-site disposal. RCRA regulates the management of soil or debris that is hazardous waste. The project proponent would need to obtain an EPA Generator Identification Number to dispose of RCRA-regulated materials (see the section on Regulatory Framework). Currently, federal regulations exempt oil- or fuel-impacted UST cleanup debris from the hazardous waste rules, but some states make no distinction between regulating petroleum-contaminated soils and other hazardous wastes.

If soil or groundwater cleanup is being performed, the definition of *clean* is a key question. There are few established federal cleanup standards for soil or groundwater contamination. State cleanup standards vary, and even states with established contaminant cleanup levels usually treat cleanup on a case-by-case basis.

Field crews may encounter groundwater monitoring wells during demolition or razing. Field crews should take precautions to not disturb these wells, as the wells may need to remain in place for many years after site closure. Unknowing or improper demolition at the surface of a groundwater monitoring or extraction well can provide a pathway for contamination to migrate from the surface into the groundwater. Also, employees should be trained to not dispose of waste oils or solvents in drinking water or groundwater monitoring wells, exposing the contractor to significant liability for cleanup.

Asbestos-containing Materials

The presence of asbestos qualifies many construction materials as *asbestos-containing materials* (ACM), and triggers a host of regulatory, procedural, and documentation requirements. Asbestos was introduced as a component of some building materials as early as the 1920s and was in widespread use by the 1950s. ACM may be present in any of the following building materials if they were installed before 1979: ceiling tile, floor tile, tile backing, tile adhesive, sprayed-on "cottage cheese" acoustical ceilings, thermal insulation on pipes, boilers, refrigeration units, roofing or flashing materials, fire doors, and Transite® siding.

Inhaling asbestos poses the hazard of asbestosis or mesothelioma. These are diseases provoked by small asbestos fibers lodging in the lungs and causing the formation of scar tissue, sharply reducing lung capacity. Strict regulations have been adopted to prevent worker exposure to asbestos and to prevent asbestos from becoming airborne through disintegration of building materials or improper removal or demolition. The term *friable* describes asbestos that can be crushed into a powdery form (the small fibers most susceptible to lodging in lung tissue).

Federal regulations require the proper removal of asbestos before any demolition or renovation activities that would break up the material. Some asbestos regulatory requirements that apply to contractors include:

- Worker and agency notification;
- Inspector training and certification;
- Worker training;
- Personnel protection and monitoring; and
- ACM-waste handling, packaging, labeling, and disposal.

Regulations require a written asbestos removal plan and notice to the EPA and other appropriate agencies before removal. Those removing asbestos must ensure the use of protective clothing and equipment for all workers engaged in abatement. They must use proper work area isolation procedures and monitor the ambient area for airborne asbestos during abatement. They must properly handle, package, label, and dispose of asbestos-containing materials, including workers' discarded protective clothing suits and respirators. ACM wastes must be disposed of in approved waste disposal facilities. State and local standards may be even stricter than federal standards for issues such as worker protection, airborne fiber limits, definitions of ACM, or notification procedures to regulatory agencies or to the public.

Polychlorinated Biphenyls (PCBs)

In a quirk of regulatory framework, PCBs are not regulated by RCRA as hazardous wastes, but are strictly regulated under TSCA (see the section on "Regulatory Framework"). PCBs typically are found in hydraulic fluids and older electrical equipment, such as transformers and capacitors. If electrical transformers supplying a facility are located off site and are owned by a local utility, there is minimal concern to the project proponent for PCB management regulations. PCBs present troublesome environmental liabilities in the event of a fire or explosion of a PCB-containing electrical component. Heavy smoke from such a fire will contact and contaminate building materials and contents, rendering all damaged materials as potentially "hazardous wastes." This occurred shortly after a new state office building opened in Binghamton, New York. Employees were forced to evacuate the premises, leaving all furniture, files, and personal effects, and were prevented from reentering for several years.

PCBs also may be encountered in waste oils. Oil recycling companies did not segregate various waste oil streams before PCBs were known to be a hazard. This resulted in the commingling of PCB-laden transformer oils with other oils for reprocessing and distribution to customers as reclaimed oils. Discovery of waste oil tanks, spills, or impact to soil or groundwater (dating from the 1970s or before) during the course of a project also raises the concern for the presence of PCBs.

PCBs also are known to have been used in fluorescent lighting ballasts. Since PCBs were removed from the market in 1978, alternate materials have been used, and fluorescent lights are labeled as being non-PCB. If large numbers of fluorescent lights are slated for removal during a facility modification or demolition, PCB management requirements may apply.

Knowledge of the PCB status of electrical equipment through site audits or by inventory or documentation from the local utility can help to minimize liabilities in case of a spill or fire. PCB regulations could apply during site modification or demolition for the use, storage, or disposal of PCBs or PCB-containing items. Those conducting site modification or demolition also should watch out for unlabeled electrical equipment containing liquids, especially during the demolition/remodeling portion of the property life cycle. Federal standards under TSCA stipulate that materials containing PCBs in concentrations of 500 parts per million (ppm) or greater must be incinerated; 50 to 500 ppm may be landfilled; and less than 50 ppm generally is not regulated. State standards in California require incineration of materials in excess of 50 ppm PCBs.

Residual Materials and Wastes

Leftover materials and other residual wastes may be encountered as any of the following:

- Unusable process raw materials;
- Unusable laboratory chemicals;
- "Last batch" wastes from discontinued processes or operations;
- Tank bottom sludge;
- Condensate from steam cleaning; and
- Miscellaneous containers of cleaners or maintenance materials.

Cleaning residues may be encountered in facility modification or expansion. The issue of leftover materials is encountered most frequently, however, at a facility shutdown or property demolition or redeployment.

Project proponents all too frequently overlook the fact that, upon discontinuation of operations, small quantities of many different types of materials that formerly had commercial value may become designated as "wastes." If containers are unlabeled or if facility personnel are unavailable to describe the origins of waste streams, then expensive full-spectrum analytical characterization may be required. Project proponents should undertake a thorough inventory of chemicals and hazardous substances well before operational shutdown, with a focus on the nonroutine items mentioned above.

Disposal options for these materials, as with all wastes, vary. Ideally, raw materials no longer suitable for use on the site may be returned to the vendor or sold for recycling or reclamation. Waste materials with sufficient heat content may be suitable for use as a supplemental fuel in an approved boiler or industrial furnace. The beneficial reuse of leftover materials (and the corresponding economic benefit) depends upon the proper identification and segregation of these materials; the unintentional commingling of reusable materials with wastes can result in additional expense to dispose of the entire inventory.

Treating hazardous wastes to render them less hazardous or nonhazardous is also possible, but may fit the definition of "treatment" under the RCRA and expose the project proponent or contractor to unwanted permitting requirements or liabilities. Some specialized contractors have obtained appropriate permits from regulatory agencies to perform chemical treatment, fixation, or soil treatment on site. In other instances, the only answer may be to remove materials for proper off-site disposal.

RCRA regulations for hazardous waste treatment, storage, and disposal are becoming increasingly stringent (see the section on "Regulatory Framework"). Many wastes now must be treated in some form before being placed into any landfill. Hazardous *liquids*, including liquid-containing materials such as sludges, may not be placed directly into landfills. Taxing schemes have been imposed in many jurisdictions to provide market-based incentives away from landfill disposal and toward recycling, reclamation, or reuse.

Those involved in facility demolition or redeployment should be assured that parties to the transaction have adequate plans in place to properly identify and dispose of residual wastes. They should ensure that imported fill is not contaminated soil from a UST excavation or spill, metal-bearing filter coke, tank bottoms, or other materials that would contaminate the project site.

Disposal of construction rubble poses the risk of commingling hazardous wastes with nonhazardous rubble, causing the mixture to be regulated as hazardous and requiring expensive disposal. This commingling could occur inadvertently on site or by sending nonhazardous rubble to another site that is accepting hazardous or contaminated soils. These risks raise the possibility of joint and several liability for cleanups.

Contractors should be certain that the imported fill they obtain and place on site is not hazardous waste or contaminated soil. Similarly, the contractor should ensure that construction rubble remains nonhazardous on the job site and is sent off site only to locations that do not commingle contaminated soils or hazardous wastes.

SUMMARY

Virtually all types of property transactions carry with them environmental risks, whether they result from past or current operations, both on site and off site. These may be residual, ongoing, or operational, personal injury, or public policy risks. Federal, state, or local environmental regulations affect essentially all transactions. Regulations continue to be a moving target, changing requirements and exposure on a continuing basis. Environmental or hazardous materials management aspects that are not specifically regulated demand good management procedures to minimize overall risks and hazards to workers, the public, and the environment.

In spite of these seemingly formidable obstacles, parties involved in property transactions can identify, minimize, and manage risks. Buyers, sellers, attorneys, and financial institutions should not blindly assume that others have addressed adequately all environmental, health, and safety issues. Parties to property transactions may develop resources in house, rely on proficient outside consultant expertise, or some combination of both options.

Environmental due diligence investigations are critical at every stage of the property life cycle. The focus of due diligence efforts may differ, depending on whether the transaction involves acquiring or developing raw land, expanding or modifying an existing facility, or demolishing or razing a site to prepare it for disposition.

While environmental issues may be the *key* issues in chemical plant and refinery transactions, they need not be "deal killers." Addressing environmental, health, and safety issues properly and early can help buyers, sellers, and other parties to minimize the risks involved and to manage liabilities. Through environmental due diligence, those involved in property transactions can make informed business decisions. In particular, buyers, sellers, attorneys, and financial institutions can use the findings, recommendations, and cost estimates that result from Phase 1 and Phase 2 assessments to properly structure the deal. The Phase 1 and Phase 2 results can guide parties to the transaction in negotiating indemnification language, preparing an agreement on shared liabilities, negotiating the price of the property, setting aside reserves to fund environmental activities, or excluding certian sites or tracts.

References

Suro, Robert. 1989. Grass-roots groups show power battling pollution close to home. *New York Times,* 2 July 1989, pp. 1, 18.

JAWMA. 1991. EPA announces FY 1991 Superfund accomplishments. *J. Air Waste Management Association*, January 1991, p. 12.

JAWMA. 1991. EPA announces reforms designed to revitalize Superfund program. *J. Air Waste Management Association*, December 1991, p. 1560.

JAWMA. 1992. EPA announces FY 1991 Superfund accomplishments. *J. Air Waste Management Association,* January 1992, p. 12.

12

Environmental Audits at Refineries and Petrochemical Facilities

David A. Dougall and Jill Barson Gilbert

INTRODUCTION

This chapter considers environmental audits conducted at refineries and petrochemical facilities. The configuration and complexity of petroleum refineries vary considerably. Generally, refineries produce a variety of fuels, intermediate petroleum derivatives and lubricants by *distillation, cracking, and/or reforming crude oil.* Petrochemical facilities process intermediate petroleum derivatives into a wide variety of organic chemicals.

Equipment commonly found at refineries and petrochemical facilities may be grouped into the following categories:

- Storage equipment—Storage of raw materials and intermediate and final products is commonly in tanks, ranging in size from a few hundred gallons to several hundred thousand gallons. Tanks may be constructed to operate under positive pressure or at atmospheric pressure. Raw materials are also commonly received and stored in metal or plastic drums or cubicles (tote bins). Dry chemicals are received in bulk and stored in silos or in containers.
- Process units—Processing generally takes place in a variety of distillation towers, reactors, and dryers.
- Utilities—Supporting equipment may consist of cooling towers, water conditioning equipment, steam boilers, heaters, wastewater handling facilities, waste storage and disposal facilities, and incinerators.

OBJECTIVES

Environmental audits are performed at refineries and petrochemical facilities for a number of reasons, each with a particular perspective. Those requesting the audit and the audit team itself should understand the objectives of the audit to insure the proper focus and reporting emphasis. Environmental audits may be conducted for lending institutions for mergers and acquisitions, for prospective buyers, for insurance underwriters, or for present owner/operators.

A lending institution or investor may request an environmental audit before financing an acquisition or refinancing for an existing owner. A pre-acquisition or pre-financing audit primarily concerns potential environmental risks (e.g., contamination) that translate into financial liabilities. A pre-acquisition or pre-financing audit for a merger, acquisition, or leveraged buyout commonly focuses on risks, but may cover those regulatory compliance issues that are of a magnitude that they may affect the continued operation of the facility.

A prospective buyer may request a pre-acquisition environmental audit. This type of audit is a comprehensive review of the environmental issues, focusing on potential liabilities (contamination, etc.) and also on operating/compliance issues that the new operator will face upon buying the facilities or in the foreseeable future. An audit conducted for a potential new operator should focus on the adequacy of systems in place to ensure environmental compliance and to minimize risk.

Insurance underwriters may require environmental audits. For these audits, the audit team must thoroughly understand the nature of the policy before structuring the audit. Environmental-related insurance frequently is written only for third-party impacts. An environmental audit conducted for this purpose should focus upon potential risks associated with gradual occurrences and sudden and accidental occurrences.

Finally, a present owner/operator may require an environmental audit, ranging from an overview of the effectiveness of environmental programs already in place to a detailed environmental compliance audit.

In general use, people apply the term *environmental audit* to compliance reviews, whereas *environmental risk assessment* focuses more on risks and liabilities and less on detailed compliance issues. Since this chapter covers the full range of environmental audits and environmental risk assessments, the term *environmental audit* is used in a global manner and *compliance audit* or *risk assessment* is used in a more specific sense. Environmental audit reports are tailored to reflect the objectives of the review. Environmental risk assessment reports typically take the form of specific findings to summarize the issues of concern. A tabulation of estimated financial liabilities is often also provided. This summary information is then followed by a more detailed report on the operations of the facilities, management programs, and a discussion of environmental issues. This type of report is particularly suited to investors and management decision makers, whose primary interest is limited to the most important issues.

Environmental compliance audit reports serve primarily as a management tool. These reports also present the audit findings up front; however, since these findings most often deal with program improvement and compliance issues, the findings frequently are prioritized in categories or tiers, reflecting a degree of concern or importance. The findings then are followed by a detailed report.

APPROACH

Organization

To be successful, an environmental audit must be adequately planned and conducted with a team of experienced personnel who are trained in environmental auditing techniques. The composition and the number of individuals on an audit team will depend upon the objectives of the audit and the size and complexity of the facilities being reviewed. In the case of an environmental risk assessment conducted on behalf of a financing client, the team may consist of a single, experienced person to conduct the site visit and plant interviews and a second person to interview appropriate regulatory personnel. A detailed environmental compliance audit on the other hand may require several members in order to provide an appropriate mix of operating knowledge and specialized regulatory compliance expertise.

The review team may be drawn entirely from an outside environmental consulting firm, as is frequently the case in environmental risk assessments conducted for financial institutions, or may be entirely internal, as is frequently the case and detailed environmental compliance audits. Outside consultants are often used in environmental compliance audits to provide experienced oversight, to lend a different perspective than company employees, or simply to supplement available experienced staff. Team selection, management approval of selected individuals, and commitment of the team members are essential parts of the audit planning process.

Training

Audit team members must receive adequate training to ensure the success of an environmental audit. In the case of outside environmental consultants, the person who commissions the audit should assess the adequacy and appropriateness of experience for the types of facilities to be reviewed. For personnel drawn from inside the company, the company should provide a training program that focuses on how to conduct environmental audits. Whether outside consultants or internal staff comprise the audit team, the company should assess the appropriateness of the field and regulatory experience of individual team members.

Scheduling

Environmental audits should be scheduled far enough in advance to obtain the commitment of team members and the acceptance of the facility being reviewed. In the case of environmental risk assessments, advance scheduling needs to provide only enough time for the facilities to make the necessary files and personnel available and to avoid conflicts with any major operational events (e.g., plant turnarounds). Detailed compliance audits, on the other hand, may require up to a week at the facility and therefore require more advance planning and time commitments by both team members and plant personnel.

Environmental audit programs that include reviews at several facilities over the course of a calendar year generally establish a 6- or 12-month schedule, which is reviewed and approved by appropriate management. This schedule then is updated as needed and circulated to appropriate personnel.

Plant management and upper management input, support, and approval of the schedule are essential ingredients in scheduling environmental audits.

Information Gathering

Information gathering for an environmental audit may include a review of documents available from the plant, corporate offices, and regulatory agencies, personal or telephone interviews with persons knowledgeable about past and present operating procedures and events at the plant, and a visual inspection of the plant equipment and property. Occasionally, if time permits and if the party commissioning the audit needs baseline data, information gathering includes collecting and analyzing samples (air, water, and/or soil) to provide additional information.

Audit teams frequently use checklists or topic outlines to aid in gathering information. Checklists may take the form of outlines of issue areas to be discussed, questionnaires, or detailed questions to be answered and discussed.

It is useful to collect at least basic plant information before the site visit. In the case of environmental risk assessments, previsit information is generally limited to the plant location, types of operations presently and historically conducted at the site, and information on environmental permits to allow interviews and/or reviews of agency files to proceed. In the case of detailed environmental compliance audits, a collection of previsit information is generally more extensive, often utilizing a written questionnaire to be completed by the plant and corporate personnel prior to the site visit.

A critical portion of an environmental audit is a review of the applicable files and plans that document the environmental programs in place, monitoring inspection procedures and records, and correspondence relevant to environmental issues. These documents may be located at the facility being reviewed, at a management office separate from the facility, or some combination of the two.

Another valuable source of information for both risk assessment and compliance reviews is the agencies that have regulatory authority at the facility being reviewed. The depth of agency review may range from telephone interviews with responsible agency field inspectors to a detailed review of enforcement and permitting files. A review of state and federal files and status lists (especially CERCLIS) for any waste recyclers or disposal facilities utilized presently or in the past by the facilities can provide valuable information.

Site Review

A site review consists of reviewing documents available at the site, interviewing appropriate facility personnel, and inspecting facility equipment and property. Personnel selected for interviews should be those who may be helpful with respect to historic and present operations, past and present waste-handling procedures, past spill events, training, inspection, and record-keeping procedures.

The audit team conducts a site inspection to review the equipment at the facility and the condition of the property. In the case of a risk assessment, a single plant inspection is generally conducted after interviews with plant operating personnel. In the case of a detailed compliance audit, a general plant tour is usually conducted after preliminary discussions to complete the orientation and familiarization of the team. A more detailed inspection of equipment and property is conducted subsequent to individual interviews and file reviews.

Table 12-1 summarizes some of the important items that the audit team should observe during the site review.

REPORT

The format and style of the report depends on the objectives of the environmental audit and also upon the preferences of the client and the author. Environmental risk assessment reports usually reflect the judgement and

TABLE 12-1 Key Observations During Site Review

The geographic settings of the facility and its relationship to off-site features, which could be impacted by the facility's operations or accidental occurrences. Examples of such features are nearby wetlands, parks, residential areas, schools, hospitals, etc.
The topography and drainage pattern of the property.
The condition and apparent state of maintenance of facilities.
The presence and apparent condition of engineering features, such as containment dikes and equipment mentioned in contingency plants or agreements with regulatory agencies.
Odors or visible air emissions.
Presence and appearance of wastewater streams.
Visible evidence of spills or leaks in process areas, tank storage areas, loading areas, and drummed material storage areas.
Secondary containment around processing and loading areas, tank and drummed storage area, and other areas where there is potential for leaks and spills.
Potential for discharging of contaminated stormwater to public waters or neighboring property.
Correct markings on tanks and other containers of potentially hazardous materials and hazardous wastes and adequate security at these types of areas.

recommendations of the reviewer, whereas environmental compliance audits generally stay very close to a presentation and summarization of only the facts. Environmental risk assessment reports generally are prepared at the reviewer's office, whereas environmental compliance audits may be prepared, at least in draft form, at the facility being reviewed. Environmental compliance audits often undergo a legal and facility management review before the final report is issued.

The primary consideration in preparing the environmental audit report is that it should meet the needs of the people for whom it is being prepared. Although the actual format of the report should be tailored to the client's needs, the report generally contains four sections: Introduction, Summary, Environmental Affairs Management, and Body.

Introduction

The Introduction should briefly present the following information:

- The dates of the review, including the time-span at the facility being reviewed as well as any reviews at corporate or regional management facilities.
- The facilities reviewed, including a brief description of the operations conducted at those facilities.
- The type of review being conducted or the objectives of the review (pre-acquisition, loan review, regulatory compliance review, focused compliance review: air quality issues, etc.).

- Composition of the audit team.
- Names and responsibilities of the persons interviewed during the course of the assessment. Include regulatory agency and management contacts as well as facility personnel.

Summary

The Summary section should provide a review for management of all the important findings and recommendations from the assessment. The report Summary frequently is expected to be used as an executive summary or a stand-alone document. This section should include the following:

- A brief statement regarding the facilities reviewed, dates of the review, and the objectives of the assessment.
- A brief reflection on the environmental management program at the facilities.
- A set of key findings, providing a concise listing of the important points contained in the body of the report. Findings may be both positive and negative or may reflect only negative discoveries (i.e., an "exception report"), depending upon the preference of the client.
- Recommendations for improvements or resolution of situations discovered during the course of the assessment. Depending on the objectives of the assessment and the preference of the client, recommendations may or may not be included in the report.
- The findings and recommendations may be grouped into categories or presented in tiers as a way of prioritizing the findings and recommendations or reflecting a degree of significance.

Environmental Affairs Management

This section of the report provides the reviewer's assessment on the existence and adequacy of an environmental management program for the facility. Considerations include the existence of explicit environmental policies and procedures, explicit assignment of responsibilities, record management systems, and communications, awareness, and training programs.

Body

The main body of the report should include a description of the facilities, the setting, and the operations, followed by a discussion of each of the environmental issue areas applicable to the facilities being reviewed. If a checklist is used in conjunction with the review, the format of the checklist may be used as an outline for this portion of the report.

Within each environmental issue area, the body of the report should discuss current procedures in use to assure compliance and/or minimize risk and describe any potential problems. If the report includes recommendations, they should be made in the context of the applicable issue areas and then restated or paraphrased in the Summary section of the report. Similarly, the report should present all of the findings in the Summary and support each in detail in the appropriate section of the Body. Certain issues may not represent actual or potential noncompliance conditions, but may represent potentially significant liabilities or potential exposure to claims against the company. The report should identify each of these exposures along with circumstances affecting the severity of the exposure and the appropriate recommendations made to reduce these exposures. The author(s) should recognize that more than one avenue or solution may be available to resolve potential problems. The report should avoid recommendations that limit flexibility in solving problems.

The audit report should include a physical description of the facility, including the size and configuration of the property, the topography, and drainage. This section also should describe the surrounding land use and identify the immediate neighbors and the nearest residential areas; identify any potential receptors of special concern, such as nearby schools, apartment complexes, hospitals, or wildlife preserves; and provide a general description of the area climate (approximate temperature, rain temperature, and rainfall ranges). If the property is known to have been in the hands of previous owners, provide a description of operations historically occurring on the property.

A description of the geology (soil types, permeabilities, and formation thicknesses) and hydrology (aquifer proximity and depth) are valuable information. If aquifers are in the vicinity of the property, information on their use for potable water supplies and their quality and yield are particularly valuable information. Any monitoring conducted on these aquifers should also be noted.

The Body should comment on housekeeping at the facilities. Photographs of the area and selected sites within the facility may enhance the usefulness of the report; however, the audit team must obtain permission to take photographs before conducting the site visit. The report should reference documents that can support significant findings or statements, and you may provide copies of pertinent documents as attachments.

ENVIRONMENTAL ISSUES

This section describes the environmental issues to be investigated during environmental audits at refineries and petrochemical facilities. These issues are common to audits of many other types of facilities. The exact manner of dividing and describing environmental issues is a matter of the preference of the investigators and the client, the detail in which each issue is investigated depends on the objectives of the review (environmental risk assessment versus compliance audit).

Solid and Hazardous Waste Management

On-site Waste Management

The federal Resource Conservation and Recovery Act (RCRA) program regulates solid and hazardous waste management activities at refineries and petrochemical plants. The U.S. Environmental Protection Agency (EPA) administers the RCRA program and has authorized most states to manage parts or all of the program. Federal and state programs carefully define *hazardous waste* and establish management systems for tracking these wastes (manifests) and for handling these wastes by generators and transporters and facilities that treat, store, and/or dispose of hazardous wastes. Generators and transporters of hazardous waste must file an appropriate notice with the EPA or the appropriate state agency, and must obtain a unique hazardous waste identification number for the facility. Virtually all refineries and petrochemical facilities generate at least some hazardous wastes and should have a hazardous waste identification number. Some states require an additional registration or identification number.

The types of hazardous wastes commonly found in petroleum refineries include oil/water separator sludge, leaded gasoline tank bottoms, slop oil emulsion solids, dissolved air floatation (DAF) float, and heat exchanger bundle cleaning sludge. Refineries also may generate RCRA hazardous wastes, such as solvents, degreasers, wastewater treatment sludge, catalysts, and maintenance wastes. Waste oil is considered hazardous in many states, but not by federal definition.

The chemical industry can generate a myriad of hazardous wastes, depending on the products manufactured. The chemical industry, as with the refining industry, also may generate hazardous wastes from normal maintenance operations and wastewater treatment, as well as incinerator residues.

Facilities that treat, store, or dispose of hazardous wastes must file a permit application to conduct those operations. Exceptions to this requirement are facilities that generate a *small quantity* of hazardous waste (defined in the regulations) or that store hazardous wastes for a period of 90 days or less.

The types of waste treatment, storage, and disposal facilities commonly seen at refineries include hazardous waste tanks and container (drum) storage areas, surface impoundments (ponds), waste piles, land treatment areas, and landfills. The federal RCRA program also may regulate some wastewater treatment units and combustion processes, based on the materials that these systems treat or destroy.

Chemical and petrochemical plants commonly have hazardous waste treatment, storage and disposal facilities (such as tanks and container storage), surface impoundments, and landfills. They also may operate incinerators, boilers, or industrial furnaces that burn hazardous waste. In addition, these plants may operate RCRA treatment facilities. Like refineries, some chemical plant wastewater treatment facilities fall under the RCRA.

Items to investigate include:

- Hazardous waste identification number;
- Regulatory status (generate, store, treat, dispose, and status of permit application);
- System of accounting for wastes generated, characterizing wastes as hazardous/nonhazardous, and record keeping on the disposition of wastes;
- Sources of wastes at the facility;
- On-site waste management facilities—facilities used (ponds, tanks, drums, piles, etc.), treatment and minimization procedures, system for monitoring and record keeping;
- Off-site disposal—description of wastes sent off site for recycling or disposal, list of transporters, recyclers, and disposers used for waste handling, system for assuring the adequacy of waste manifesting, and system for minimizing the risks associated with contract transporters and disposers; and
- Account of inspections by regulatory agencies and their findings.

Soil and Groundwater Contamination

Petroleum refineries and petrochemical plants are large and complex facilities that may include miles of pipelines carrying raw materials, intermediate and final products, and large numbers of vessels containing various hazardous chemicals. Older refineries often have buried pipelines and unlined ditches. Historically, spills and leaks from vessels and pipelines may have soaked into the soil or may have been carried along with process wastewater through underground sewers or open ditches to containment and/or treatment facilities. Unlined pits and equalization basins have been common in the past at these facilities. Chronic leaks of wastes, raw materials, and products from tanks and pipelines have not been unusual. Burial of wastes, including sludges and leaded gasoline tank bottoms, also have been historically commonplace.

The RCRA, Superfund, and Underground Storage Tank programs all regulate releases of substances to the soil and groundwater. These programs have different reporting, investigation, and remediation requirements. Investigation and cleanup activities can be costly, so proper management is essential. In addition, civil and criminal penalties can occur from noncompliance.

For these reasons, a reviewer should be sensitive to the possibility of soil and groundwater contamination at the facility and should investigate whether a program is in place to confront the issue. The impetus for a subsurface contamination investigation often is associated with the RCRA Part B Hazardous Waste Treatment/Storage/Disposal facility permit.

Whatever the source of the soil and groundwater contamination, the audit team should review the program with the plant environmental coordinator to assure that positive steps are being taken to manage the problems and that agency-prescribed timetables or compliance schedules are being adhered to. The magnitude of the program should be understood by the reviewer, especially with respect to the presence or likelihood of migration of contamination off the property and the presence or probability of impact on freshwater aquifers that may be used for drinking or irrigation.

Off-site Cleanup Activities

Because refineries and petrochemical plants sometimes generate substantial volumes of hazardous waste that may have been shipped to off-site commercial disposal, these facilities are commonly named as potentially responsible parties (PRPs) in remediation proceedings in federal Superfund and state cleanup sites.

The audit team should review the present program at the facility intended to ensure that hazardous wastes sent off site for commercial disposal are being properly handled, to avoid creating liabilities with new Superfund sites. A reviewer also should discuss with the plant environmental coordinator any known or pending actions that will impact the facility as a PRP based on past waste management activities. In addition, the reviewer should examine the current Comprehensive Environmental Response Compensation and Liability Information System (CERCLIS) list for any disposal sites known to have been used by the facility.

Water and Wastewater Management

Potable Water

The federal Safe Drinking Water Act (SDWA), as amended in 1986, requires that the EPA set *maximum contaminant levels* (MCLs), *maximum contaminant level goals* (MCLGs), and national primary drinking water regulations for 83 specific contaminants and for any other contaminants that may have adverse effects upon public health. Monitoring requirements and applicable standards are applicable to *public water systems*; however, because of their size and complexity, many refineries and petrochemical plants will be affected by these rules.

During an environmental assessment, look for a system in place that evaluates the facility's potable water service for the applicability of drinking water rules and that provides compliance assurance for monitoring and quality criteria.

During an environmental review, the audit team should look for the following:

- Description of the potable water system, including water sources treatment, storage, and distribution;
- Interpretation of drinking water rules for applicability to the plant potable water system;
- Drinking water analytical results; and
- Documentation of compliance with monitoring and quality criteria.

Wastewater

The Federal Water Pollution Control Act (FWPCA), most recently amended by the Water Quality Act (WQA), is the principal regulatory authority for industrial wastewater. Industrial facilities with point source discharges of wastewater directly into waters of the United States or their tributaries must obtain permits issued under the National Pollutant Discharge Elimination System (NPDES) program. The EPA is the primary permitting authority, though the EPA has authorized most states to manage the program. In some states, both state and EPA permits are required, because the state is not authorized to conduct the federal program. Industrial facilities that discharge to publicly owned or regional treatment facilities, rather than directly to surface waters, must meet prescribed pretreatment standards rather than obtain an NPDES permit for the discharge.

NPDES permits describe the discharge points or *outfalls* to which they apply and prescribe specific effluent quality characteristics that the facility must meet and document by required monitoring and reporting procedures. Similarly, facilities that discharge to publicly owned or regional industrial wastewater treatment plants should have a document that prescribes discharge criteria and monitoring requirements that must be met.

Some wastewater management issues unique to refineries include rainfall within tank dikes, sour water and spent caustic systems, and API oil/water separators. Many older refineries do not have segregated process water and stormwater management systems, so they must treat large volumes of commingled wastewater. Often, older refineries and petrochemical facilities have large surface impoundments to hold stormwater before treatment. These impoundments can create soil and groundwater contamination through leaks over a period of years. Newer facilities usually have segregated systems and aboveground treatment facilities, thus reducing risks.

The key to a review of wastewater management is to ensure that the NPDES permit includes all wastewater discharges, including stormwater in many cases, that the permit is current (i.e., not expired) and issued to the current owner/operator, and that the permit conditions are being met.

The audit team should review the following information:

- Sources of wastewater (processes, cooling towers, boiler blowdown, water conditioning, sanitary sewers, stormwater, etc.);
- On-site wastewater management—Processes/facilities used, design limitations, chemical used, special problems (unlined ponds, other);
- Monitoring systems—Effluent limitations, sampling responsibilities, analytical responsibilities (internal versus contract), review, and reporting responsibilities;
- Off-site wastewater disposal—Description of disposal procedure, list of transporters and disposers, system for assuring adequacy of disposal, hazardous/nonhazardous; and
- Inspections by regulatory agencies—Account of previous inspections and agency findings.

Spill Containment and Contingency Planning

Several federal laws and corresponding regulations govern spill and incident reporting. The federal Clean Water Act prohibits the discharge of harmful quantities of oil into the surface waters and requires most facilities that store oil or oil products to prepare a Spill Prevention, Control and Countermeasure (SPCC) plan. Spills of oil that result in a sheen on any surface water are required to be reported to the National Response Center. Facilities with any of a great number of other chemicals are not presently required by regulation to prepare and implement formal spill prevention and control plans, but they are required to report spills that exceed a *reportable quantity* under the Comprehensive Environmental Response, Compensation and Liability Act (CERCLA) and also to report other releases under the Superfund Amendments and Reauthorization Act, Title III (SARA Title III).

The key to review with respect to spill containment and contingency planning is to ensure the presence and adequacy of spill response plans, personnel training, and facility design and construction. Review of the facilities should include the following:

- Spill response/contingency plans—Includes all potentially hazardous materials and wastes, comprehensive and up to date;

- Training—Employees properly trained in day-to-day handling of chemicals and in spill response and emergency procedures;
- Bulk storage facilities—Adequate design, materials of construction, high-low-level shutdown/alarms, security, secondary containment, inspection procedures, and record keeping;
- Container storage facilities—Designation of container storage areas, site security, inspection procedures, secondary containment, and drainage; and
- Loading/unloading facilities—Materials handled, operating/inspection procedures, unit drainage, and secondary containment.

Underground Storage Tanks

The Hazardous Solid Waste Amendments of 1984 (HSWA) require regulation of petroleum and hazardous substances underground storage tanks (USTs), separate for the RCRA program. The RCRA program regulates hazardous waste management in USTs. The EPA has authorized most states to conduct the federal program. Most states have additional requirements, depending on local geologic and groundwater conditions.

UST programs have a substantial impact on the underground storage of petroleum and hazardous chemicals, especially at gasoline service stations. UST programs are based on federal regulations implemented and administered by state agencies. The programs require new USTs to meet high standards of leak and spill prevention, to cause existing tanks to be upgraded or removed over the period of the next several years, and to cause leaks and spills that have occurred or are occurring to be detected and the resulting contamination cleaned up.

Regulations apply to USTs and associated underground piping that store petroleum or hazardous chemicals. The regulations exclude coverage of very small tanks (less than 110 gallons), flow-through process tanks, tanks used in wastewater treatment, and emergency overflow tanks.

The key to an environmental assessment is to assure that a program is in place for a comprehensive review of tanks and a schedule of upgrading or removal. The audit team should investigate the following items:

- UST registration forms and agency certificates, if applicable in the particular state;
- Inventory of USTs with dates of installation, materials stored, condition, and so on;
- Schedule for removal/upgrade;
- Inventory reconciliation program or leak detection equipment in place;
- Appropriate procedure for removal of USTs;
- Documentation of *clean closure* for tanks already removed; and

- Documentation of any UST integrity tests, cathodic protection tests, and contamination investigations.

Air Emissions Management

The federal Clean Air Act, originally passed in 1970 and amended in 1977 and 1990, regulates the air emissions at refineries and petrochemical plants. The EPA authorizes most states to manage the federal programs. Most states also have air pollution control laws that can be more stringent than federal programs. In some states, regional or local agencies administer the air quality programs. The remainder of this discusses key federal air programs and issues specific to refineries and petrochemical plants.

National Ambient Air Quality Standards

The National Ambient Air Quality Standards (NAAQS) establish acceptable ambient levels for specified *criteria* pollutants (particulate matter, carbon monoxide, nitrogen oxides, sulfur dioxide, ozone, and lead). Geographic areas are classified as either *attainment* or *non-attainment*, which means that they either meet or do not meet the standards, respectively. States that do not meet the NAAQS must prepare state implementation plans (SIPs) to demonstrate progress towards attaining and maintaining NAAQS. These plans can have great impacts on the ability of refineries and petrochemical plants to operate in non-attainment areas.

New Source Performance Standards

State and EPA permit programs are designed as the primary mechanism in conducting the air quality control program. Through these permits, the agencies can require use of technologies to minimize emissions. Several types of regulations determine emission limits, controls, and operating procedures in these permits. One major type of regulation is the New Source Performance Standards (NSPS). Several NSPS apply specifically to refineries or the synthetic organic chemical manufacturing industry (SOCMI) and to equipment often found at these facilities.

The NSPS apply to specific new or modified sources. Several NSPS affect units at refineries and petrochemical plants. Some NSPS apply only to units at refineries or facilities within the synthetic organic chemical manufacturing industry (SOCMI); others apply to units in general industrial use, including refineries and petrochemical plants. NSPS apply only to units meeting prescribed criteria, which include the date of construction or modification and may include other factors, such as size, throughput, vapor pressure of materials handled, and so on. These standards set specific emission limitations and prescribed monitoring, record keeping, and reporting.

The applicability of NSPS usually is acknowledged in preconstruction permits, as well as the emission limits and monitoring, record keeping, and reporting incorporated into the permit conditions. However, because NSPS apply to affected facilities modified or constructed after the date of *proposal* of the regulation, but did not take effect until *promulgation* of the regulation, construction permits occassionally have been issued for facilities or modifications between the dates of proposal of an applicable NSPS and its promulgation. As a result, the requirements of the applicable NSPS have been omitted from the permit. With this in mind, the audit team should scrutinize sources at refineries and petrochemical plants for which NSPS have been proposed or promulgated to determine whether the NSPS applies and whether a system is in place to ensure compliance.

The following NSPS may apply to refineries or petrochemical plants. Audit team members responsible for air quality issues should become familiar with the subparts potentially applicable to the facility being reviewed.

- Subparts Da, Db, and Dc–Steam Generators;
- Subpart J–Petroleum Refineries;
- Subparts K, Ka, and Kb–Storage Tanks;
- Subpart GG–Stationary Gas Turbines;
- Subpart VV–Volatile Organic Compounds (VOCs) at SOCMI;
- Subpart GGG–VOCs at Refineries;
- Subpart III–SOCMI Air Oxidation Units;
- Subpart NNN–SOCMI Distillation Operations; and
- Subpart QQQ–Petroleum Refinery Wastewater Emissions.

National Emissions Standards for Hazardous Air Pollutants

Another major aspect of the overall clean air program is National Emission Standards for Hazardous Air Pollutants (NESHAPs). Whereas NSPS are technology-based and aim to set specific emission limits, NESHAPs are procedural in nature and aim to minimize emissions by establishing operating practices.

Benzene is one of several compounds for which federal regulations have been promulgated for controlling emissions. The benzene NESHAPs apply to existing sources, as well as to new and modified facilities. Benzene emissions are controlled by several different NESHAPs, each applicable to a different type of potential source. Benzene NESHAPs that may apply to refineries and petrochemical plants are as follows:

40 CFR Subpart J–Equipment Leaks (fugitive emissions sources) of Benzene—This regulation applies to certain types of equipment (pumps, com-

pressors, valves, flanges, etc.) that are *in benzene service* (i.e., contact liquid or gas streams with greater than 10% benzene by weight). Affected facilities must conduct a fugitive emissions monitoring and control program that includes regular testing of potential fugitive emissions and repair of sources discovered to be leaking in excess of the allowable amount.

61 CFR Subpart Y–Benzene Storage Vessels—This subpart applies to tanks (over 10,000 gallons) used to store industrial grade or refined benzene. These tanks must be of floating roof construction or have a close-vent system with control device, and the tanks must comply with prescribed seal requirements. Testing and record keeping are also prescribed.

61 CFR Subpart BB–Benzene Transfer Operations—This subpart applies to loading racks at production facilities and bulk terminals loading benzene (at least 70% by weight) onto trucks, railcars, or marine vessels. This subpart does not apply to racks loading only gasoline, to liquid wastes containing benzene (covered under Subpart FF), or to coke by-product recovery plants (covered under Subpart L). Affected facilities must provide vapor collection and control, with 98% reduction of benzene emissions and other procedures to minimize and test for leaks. Testing and record keeping is also required.

Subpart FF–Benzene Waste Operations—This subpart applies to facilities with treatment and control and devices used to manage benzene-containing waste streams, including wastewater treatment and other operations. Benzene-contaminated waste streams must be treated to reduce benzene concentrations to prescribed limits, and units must be equipped with covers to prevent detectable emissions. Monitoring and record keeping are required.

The key to conducting a review of facilities that may handle benzene or benzene-containing wastes or byproducts is for the facility person responsible for air quality issues to describe their *system* for evaluating the applicability NESHAPs to their operations for determining the units and processes to which the standards apply and for documenting compliance with design monitoring and record-keeping requirements.

Hazardous Air Pollutants

While some toxic compounds are limited through NSPS and NESHAPs, the 1990 Clean Air Act (CAA) Amendments add substantially to the level of air toxics control. Title III of the act lists 189 *hazardous air pollutants* (HAPs) and 16 extremely hazardous substances. Facilities that emit more than 10 tons per year of any single HAP or 25 tons per year of combined HAPs will be required to install advanced controls. Standards for these controls are expected to be issued through November 1992. Facilities with greater than threshold quantities of *extremely hazardous substances* will be required to develop risk management plans that incorporate detection and minimization of accidental releases.

Odor

There are no federal regulations that specifically require the control of odors from industrial facilities, and state controls generally are limited to nuisance laws. Nonetheless, odors from refineries and petrochemical plants sometimes create problems that result in chronic complaints from the neighboring community.

A reviewer should determine whether odor complaints are a problem at the facility and should be alert to odors at the facility throughout the review. Regular employees at the plant may become accustomed to odors at the facility and miss opportunities to correct nuisance odors that are apparent to a visitor.

Permits

Given their size and complexity, many refineries and petrochemical plants have a substantial number of permits and amendments and also a number of units that are *grandfathered,* that predate or are otherwise exempt from permitting requirements. The environmental audit team should look for a comprehensive system that presents an organized description of potential emission sources and that documents how they are authorized (grandfathered, exempt, permitted) and how compliance with permit conditions is assured.

For an environmental review, consider the following:

- Sources permitted;
- Permit conditions known;
- Testing/monitoring conducted and documented;
- Compliance history;
- Results of agency inspections; and
- Complaints from neighbors, community.

Polychlorinated Biphenyls (PCBs)

The federal Toxic Substances Control Act (TSCA) regulates PCBs and PCB-contaminated materials. PCBs historically have had substantial use in the electrical systems of refineries and petrochemical plants because of their improved fire safety characteristics. By the PCB rules, electrical equipment (transformers, capacitors, and rectifiers) containing fluid with greater than 500 parts per million (ppm) PCB must be inspected at least once per month, with cleanup and repair of any leakage. The facility must prepare an annual report that documents the quantities of PCBs on hand at the beginning and end of each calendar year, showing the disposition of any PCBs during the year. Items and equipment containing PCBs greater than 500 ppm must bear special labels, and facilities used to store PCBs must meet specified criteria.

In addition to electrical equipment, PCB contamination has been found in some compressor lubricants and heat transfer fluids that could conceivably be present at refineries or petrochemical plants. Spills of PCB must be reported and cleaned up in accordance with specified criteria.

During an environmental assessment, look for a system that provides an inventory of all PCBs and PCB-contaminated equipment or soil and that indicates an awareness of monitoring and reporting requirements.

The audit team should review the following:

- Annual PCB inventory;
- Documentation of testing for reclassified transformers;
- Manifests for PCBs disposed of;
- Appropriate markings on PCB equipment, PCB liquid containers, PCB transformers, and storage facilities;
- PCB storage accommodations in accordance with requirements; and
- Record of spills, notifications, and cleanup.

Asbestos Management

Asbestos is not considered a hazardous waste under RCRA; however, because of health concerns with airborne asbestos fibers, it is considered an HAP. Federal NESHAPs regulate the management of this material (61 CFR Subpart M). Provisions of this subpart primarily impact refineries and petrochemical plants, where work is planned that will result in stripping or demolishing equipment or piping that is insulated with asbestos-containing materials. Although insulating materials are the primary potential source of asbestos-containing materials at refineries and petrochemical plants, asbestos is also sometimes found in some asphalt floor tile and tile adhesives, some gaskets, brake linings, and some Transite® wallboard.

With limited provisions for excluding small jobs, advanced notice must be filed with the EPA and/or the appropriate state agency before undertaking any removal or demolition activities that involve asbestos-containing materials. Prescribed procedures must be followed for removing and holding asbestos-containing materials (primarily wetting), and disposal must be in a landfill approved for receiving asbestos. Use of an RCRA hazardous waste landfill is not required for disposal of asbestos-containing materials.

When reviewing a facility for proper asbestos handling, determine whether the facility has prepared an asbestos management plan (a valuable management tool, but not a regulatory requirement). Determine whether the facility has been reviewed for the presence of asbestos-containing materials and the locations mapped, marked, or otherwise designated. Look for an internal procedure that assures that activities that might involve removal or demolition of asbestos-containing materials comply with notification, removal, holding, and disposal requirements. Also, look for a system to assure that contractors not only perform to job specifications at the plant, but are properly trained and licensed and will account for proper transportation and disposal of all wastes generated.

Pesticides Management

The Federal Insecticide, Fungicide, and Rodenticide Act (FIFRA) regulates the registration, manufacture, transportation, use, and disposal of pesticides. Manufacturers of pesticides must be registered with the EPA. Pesticide operations are subject to reporting and labeling requirements.

During an environmental review, look for the following:

- Pesticide application contractors are licensed.
- Pesticides applied by plant personnel have EPA and registration number on label.
- Pesticide containers are properly disposed of.

Community-Right-to-Know (SARA Title III)

The Emergency Planning and Community Right-to-Know Act of 1986 is Title III of the Superfund Amendments and Reauthorization Act of 1986 (SARA). Although this Title is within SARA, Title III is entirely separate from the Superfund program. Facilities are required to submit information on the amounts of chemicals used, manufactured, and released. A key feature is that most of the data submitted is subject to public disclosure and has increased community concerns about the releases from facility operations.

The EPA administers national implementation, with state emergency response authorities designated by the governor of each state for emergency planning and response coordination. The management of a facility should be aware of the emergency planning process and status, and may have a representative on the local committee.

The EPA published a list of *extremely hazardous substances* (EHS), which has been revised several times since it was initially issued. It presently lists about 360 chemicals. Each chemical on the list has a *threshold planning quantity* (TPQ), which ranges from 10,000 pounds to as little as 1 pound. A facility that manufactures, imports, processes, or otherwise uses more than the TPQ is required to provide applicable material safety data sheets (MSDS) or a listing of chemicals handled to the state and local emergency planning agencies, and must designate a SARA Title III Facility Emergency Coordinator. The initial information submittal was due by May 17, 1987.

Under Sections 311 and 312 of SARA, facilities must submit annual chemical inventory forms (Tier One or Tier Two Reports) if the amount of chemicals at the plant exceed prescribed criteria. Additionally, a SARA Section 313 annual Toxic Chemical Release Inventory (Form R) must be submitted if one or more of the listed chemicals are manufactured, imported, processed, or otherwise used above the applicable threshold quantities. The list includes over 300 chemicals and several groups of chemical compounds. A release is defined as *any spilling, leaking, pumping, pouring, emitting, emptying, discharging, injecting, escaping, leaching, dumping, or disposing into the environment (including the abandonment or discarding of barrels, containers, and other closed receptacles) of any hazardous chemical, extremely hazardous chemical, or toxic chemical.*

During an environmental review, look for the following:

- A system in place that evaluates the applicability of SARA Title III to the plant and inventories chemicals to which it applies.
- Tier One or Tier Two Reports and Form Rs and supporting documentation.
- Release response procedures, including both internal and external activities.

SUMMARY

Refineries and petrochemical plants often are complex facilities with a variety of processes. Several types of parties may request environmental audits: lenders, buyers, insurance underwriters, and current owners and operators. These parties may request audits to determine risks, to determine the facility's compliance status, to predict future expenditures, to gauge the facility's value, or for other purposes.

A successful environmental audit requires trained personnel who are experienced with refinery and petrochemical operations and knowledgeable of regulatory requirements and risk issues. Facility management cooperation is essential to scheduling and conducting the data review and site visit.

The audit team should understand the purpose of the audit and direct the information gathering and reporting activities accordingly. Environmental audit reports vary, depending on the audience. Some contain only findings;

some contain recommendations; others focus on cost projections for environmental issues.

Refineries and petrochemical facilities have many diverse environmental issues that cover not only air emissions, but also water and waste management and numerous other areas. Some of the air issues overlap other disciplines, highlighting the importance of a knowledgeable, well-trained audit team.

13

An Audit Approach to Source Reduction of Air Toxics and Ozone-Depleting Chemicals

Pamela Beilke and Linda Micale

INTRODUCTION

Pollution prevention is a fairly new environmental program that focuses on the elimination of waste through source reduction and recycling rather than through treatment and disposal. Driving forces behind the program include the high cost of effectively treating and disposing of wastes, the diminishing capacity of existing facilities to safely handle waste treatment or disposal, and the soundness and expediency of a preventative approach. The Federal Pollution Prevention Act, passed in 1990, requires facilities subject to toxic chemical release reporting under Section 313 of the Emergency Planning and Community Right-to-Know Act (EPCRA) to submit an annual report on source reduction and recycling of listed toxic chemicals.

This emergent policy has broad implications for a number of other programs administered by the Environmental Protection Agency (EPA). Regulations promulgated under the Resource Conservation and Recovery Act (RCRA) and the Hazardous and Solid Waste Amendments (HSWA) are expected to be altered, in large part, on the basis of our nation's move towards the prevention of pollution. For example, the EPA formally requested comment from industries on the ways in which hazardous waste regulations under the RCRA may hamper the implementation of pollution prevention programs. Trade journals are replete with examples of reductions in RCRA enforcement penalties, in favor of the facility installing pollution prevention programs or control equipment. Recently, stormwater or non-point source runoff has been hailed as the nation's most difficult water quality challenge. Stormwater permitting now required under the Clean Water Act regulations will encompass the use of best management practices aimed at preventing or controlling runoff. The correlation between EPCRA and pollution prevention has been duly noted. EPCRA Section 313 reports offer the means by which the EPA will initially identify and track facilities subject to requirements and having the potential for reduction or prevention of pollution.

The increasing regulation of air emissions provides a salient example of the need for identifying and implementing pollution prevention opportunities. Concerns over the high cost and difficulty of compliance have arisen among industries in light of new standards for air emissions from RCRA facilities, reporting of toxic chemical release reductions under the Federal Pollution Prevention Act of 1990, and regulation of air toxics and ozone-depleting chemicals under the Clean Air Act (CAA) Amendments. Title III (Hazardous Air Pollutants "Air Toxics") and Title VI (Stratospheric Ozone Protection) of the CAA Amendments of 1990 will also have a significant impact on industrial sources. This chapter will review pollution prevention concepts, laws, and programs, with special emphasis on reducing air emissions. The use of auditing tools will be addressed to assist industries in identifying substantial opportunities to reduce or eliminate emissions of air toxics and ozone-depleting chemicals.

OVERVIEW OF POLLUTION PREVENTION PROGRAMS

Pollution prevention programs fall under two basic categories: waste minimization and source reduction. Source reduction is preferred by the EPA because it involves the identifying and subsequently reducing chemicals used in a industrial process. Typically, source reduction efforts are "front-end" activities involving inventories of targetted chemicals to record purchasing patterns, their quantities, where and how they are stored, and the hazards that they pose. Evaluations of processes that use these chemicals are undertaken to consider the potential for suitable substitutes that result in safer and lower volume chemical storage requirements. Waste minimization, by contrast, focuses on the effectiveness of process applications, proper characterization and segregation of wastes, and the potential for reuse, reclamation, or recycling, to result in a waste of lower toxicity or volume. In this case, pollution prevention efforts target end results (i.e., waste-generating activities).

In the Pollution Prevention Act (PL 101-508), Congress declared that it will "be the national policy of the United States that pollution **should be prevented or reduced at the source whenever feasible;** pollution that cannot be prevented should be recycled in an environmentally safe manner, whenever feasible; pollution that cannot be prevented or recycled should be treated in an environmentally safe manner whenever feasible; and disposal or other release into the environment should be employed only as a last resort and should be conducted in an environmentally safe manner."

THE FEDERAL PROGRAM—TOXIC CHEMICAL SOURCE REDUCTION AND RECYCLING

As noted, the federal program currently centers on the concise Pollution Prevent Act of 1990. This primarily outlines goals for approaching the nation's hazardous substance and waste management issues in a progressive, rather than reactive, manner. The solitary requirement of the Act is having to submit a Toxic Chemical Source Reduction and Recycling Report (TCSRR) to the EPA if your facility already files under EPCRA Section 313. The first annual report was due on July 1, 1992.

The contents of the TCSRR are summarized below:

- Quantity of chemical entering any waste stream or released to the environment before recycling, treatment, or disposal, and the percentage change in the preceding year;
- Amount of chemical recycled (on site and off site), percentage change, and process used;
- Source reduction used, including equipment or process modification, reformulation or redesign, substitution, improved housekeeping;
- Estimate of amount to be used and recycled for the next 2 years and percentage change from reporting year;
- Ratio of production in reporting year to the previous year;
- Techniques used to identify source reduction;
- Amount of toxic chemical released due to catastrophic event, remedial action, or other episodic release; and
- Amount of chemical treated (on site and off site) and percentage change from previous year.

This reporting requirement is limited to those toxic chemicals reported under Section 313. The TCSRR report will be filed in conjunction with the Section 313 report on toxic substances releases.

In light of the link made between EPCRA reporting and the TCSRR, it appears clear that the EPA seeks data at this time, potentially on which to base future regulations. For the present, the TCSRR may be utilized as a self-regulating tool by astute facilities. A practical observation of the implementation of EPCRA 313 is that industries learned more about their own systems by completing this reporting process, became subject to a certain level of public scrutiny, and made prompt and voluntary decisions to reduce levels associated with the toxic substances release report. It is quite possible that the TCSRR provides reporting facilities with the means to put a context on their release reporting numbers and demonstrate to the public that there are many systems in place to prevent and mitigate pollution.

The Act is one of many vehicles by which federal and state governments are targeting preventative measures to eliminate pollution. In the last few years, a significant number of waste management bills introduced to Congress bear the banner of source reduction, recycling, or waste minimization. States like New York, Oregon, and California have legislated pollution prevention programs that carry more strict reporting requirements, contain penalties, and require pollution prevention plan to be submitted to ensure that a regulated facility incorporates such efforts into their basic operations.

A State Example—The Arizona Program

The Arizona program is noteworthy because it combines reporting and planning at the facility level with a requirement for a progress report to the Arizona Legislature by the Arizona Department of Environmental Quality (ADEQ). Even the regulatory agency is being examined on its success in this arena. By January 1, 1994, the ADEQ must establish a numeric statewide goal for waste minimization, followed in 1999 by a numeric goal for toxic use reduction.

Arizona's reporting requirement currently extends beyond the 313 reporting facilities to all large-quantity generators of hazardous wastes or to those facilities generating more than 1 kilogram of acutely hazardous waste per month during the calendar year. The Arizona program also requires submittal of a Pollution Prevention Plan prepared in accordance with A.R.S. Title 49, Article 4, Section 49-963.

The Pollution Prevention Plan must address hazardous wastes and toxic substances. The statute defines toxic substances as hazardous substances under the state (A.R.S. 49-281) and federal (Section 101 or 102 of CERCLA) Superfund, toxic chemicals under Section 313 and extremely hazardous substances under Section 302 of EPCRA, imminently hazardous chemicals under Section 7 of the Toxic Substances Control Act, and substances listed as a known or reasonably anticipated to be carcinogen, pursuant to 42 USC 241(b)(4).

The plan must cover a period of at least 2 years and must include the following:

- Name, location, and principal business activities;
- Name, address, telephone number of owner or operator, and senior official;

- Certification by senior official that he or she "has read the plan and it is to the best of my knowledge true, accurate and complete";
- Specific performance goals for the facility and explanation of rationale for each goal, which may include individual processes;
- Written policy setting forth management and corporate support and commitment to implement the plan to achieve the goals;
- Statement of scope and objective;
- Analysis of opportunities to reduce or eliminate toxic substance releases and hazardous waste generation;
- Analysis of pollution prevention activities already underway;
- Employee awareness and training programs;
- Provisions to incorporate the plan elements into management practices and procedures; and
- Description of options considered and explanation of why options were not implemented.

An annual progress report, submitted to ADEQ on July 1 of each year, must analyze progress made relative to each performance goal, amend the Pollution Prevention Plan as needed, and explain the need for amendments. The first plans were due on December 31, 1992; facilities that ship from 10,000 to 12,000 kilograms of hazardous waste off site for purposes other than recycling must also submit a plan to ADEQ by December 31, 1995.

A CLOSER LOOK AT AIR TOXICS AND OZONE-DEPLETING CHEMICALS

Air toxics and ozone-depleting chemicals are prime candidates for source reduction, not only as a result of the EPA trend towards pollution prevention, but also due to the extension regulation of these chemicals under air regulatory programs.

Air Toxics Regulations

Prior to the 1990 amendments, the EPA established toxic air pollutant standards on a health risk basis. Now the EPA must specify maximum achievable control technologies (MACT) for 189 "air toxics" listed by Congress in the CAA amendments. If a facility emits more than 10 tons of any single air toxic or more than 25 tons of any combination of air toxics in a year, MACT requirements will apply. The EPA can also apply MACT standards to area sources, such as gas stations, dry cleaners, and wood burning stoves.

The emission control standards must be established for source categories, rather than specific chemicals. The standard will identify the technology that must be utilized by new and existing sources. The EPA may, however, allow product substitution, process modification, and improved work practice as an alternative to end-of-the-pipe controls.

Congress established a schedule for promulgation of MACT standards for these source categories as follows:

- November 15, 1992—First 40% MACT standards;
- November 15, 1994—First 25% MACT standards;
- November 15, 1997—Second 25% MACT standards; and
- November 15, 2000—Final 50% MACT standards.

The EPA was required to publish its final list of source categories and subcategories on November 12, 1991. The draft list was issued by the EPA on June 21, 1991. The EPA was required to publish a schedule for promulgation of MACT standards for each source category by November 15, 1992. Once promulgated, existing industry sources will have 3 years to meet the MACT standard.

Operating permits are also required for major sources of air toxics. The EPA promulgated final regulations establishing a federal operating permit program on November 14, 1991. The permitting program will actually be implemented at the state and local level; therefore, state agencies must adopt adequate legislation and regulations before implementing the program.

Operating permit applications, which must be submitted by November 15, 1995, must include:

- General facility information—name, address, owner, operator, facility contact;
- Description of facility processes and products, SIC code;
- Inventory of emissions of *all* regulated pollutants;
- Description of all emission points and emission rates;
- Description of fuels and raw materials and how they are used;
- Description of pollution control equipment;
- Description of operating limitations or restrictions on work practices that affect emissions;
- Description of applicable federal and state control requirements; and
- Description of test methods used to determine compliance.

The EPA was also required to issue a list of extremely hazardous substances and associated thresholds for preventing accidental releases by November 15, 1992. If a facility has quantities of listed materials in excess of the threshold, the facility must establish a risk management plan that provides a history of past releases and that defines an accidental release prevention program, a contingency plan for release response, and public information procedures.

Ozone-Depleting Chemicals

The Montreal Protocol, signed in 1987, established a global commitment to eliminate the production of chemicals that contribute to the depletion of stratospheric ozone. In Title VI of the CAA Amendments, the United States formally adopted a schedule for the phasing out chlorofluorocarbons (CFCs), halons, carbon tetrachlorides, and methyl chloroform. This schedule is more aggressive than that originally adopted by the Montreal Protocol.

As production of these chemicals is phased out, the cost continually rises. Industrial users must aggressively pursue alternatives to using ozone-depleting chemicals.

AN AUDIT APPROACH TO SOURCE REDUCTION OF AIR TOXICS AND OZONE-DEPLETING CHEMICALS

An audit is a systematic process that is applicable to a variety of topics. Audits can be broken down into four main elements: preaudit activities, field audit, report generation, and follow-up.

The preaudit, often called a "desk audit,"includes defining the objective and scope of the evaluation, selecting the evaluation team, preparing the evaluation materials, completing a regulatory update, and intensively evaluating the written procedures used at the facility.

The field audit activities include an opening conference, careful review of documentation, inspection, interviews, and issuance of corrective action forms. A closing conference should be held following the completion of the field audit to highlight the findings of the evaluation.

The final report should be generated as quickly as possible and should include an introduction, a summary, audit findings and concerns, recommendations, a conclusion, and corrective action status.

This basic process can be applied to the evaluation of opportunities for source reduction of air emissions. The objective of this type of audit is to characterize emission sources of targeted chemicals and to identify opportunities for reducing or eliminating these emissions. The basic steps are:

1. Identify chemicals containing air toxics and ozone-depleting chemicals;
2. Assess existing emission sources;
3. Estimate emission rates;
4. Identify emission reduction alternatives; and
5. Develop an action plan.

In addition to characterizing emissions and facilities at the process level, a key component to source reduction of air toxics involves evaluating the chemical purchasing and distribution practices at a facility. Questions to consider include:

- Are departments using two distinct chemicals for the same purpose?
- Is one chemical more toxic than the other or result in exceedance of a reporting threshold?
- How is chemical authorization occurring; is regulation, reportability, or potential for source reduction of chemicals being factored into purchasing decisions? In this way, the air toxics audit would also focus on the front-end activities more closely associated with reducing emissions at the "source," or prior to generation of any physical waste or emission.

Preaudit Activities

The audit team should include members with expertise in regulatory requirements, process engineering, and toxicology. The preaudit begins with a review of Material Safety Data Sheets (MSDS) for chemicals used at the facility, to identify those that contain target chemicals. In this review, a number of information sources should be consulted, including the facility hazardous materials inventory, EPCRA Tier 2 and 313 reports, annual air emission reports, and procurement records.

The regulatory review includes identifying all sources currently subject to federal, state, and local regulatory and permitting requirements. These requirements may already include emission limits and technological controls.

It is important in the preaudit to collect, review, and compile information regarding existing air emission sources, including air permits, facility floor plans, process flow diagrams, equipment information on rated capacity, and pollution control equipment. In addition to these data, the auditor must become thoroughly familiary with hazardous materials procurement and distribution practices. The auditor must develop checklists tailored for air and source reduction purposes to be used during the field audit to record upfront materials management processes and to verify the location of permitted equipment, chemicals usage rates, and existing control equipment and technologies.

An intensive evaluation of the written procedures used at the facility will provide information regarding chemical selection and procurement, equipment operation and maintenance, and emission monitoring.

Field Audit

The field audit should verify compliance with existing requirements, verify permitted equipment, and ensure that permit requirements are being met. It is important to look for non-permitted equipment that should be permitted, evaluate intermittent and non-process activities, and verify proper operation and maintenance, such as monitoring for leaks and prompt repairs.

During this review, it is important to try to identify each device or process that releases target chemicals, including process vents, equipment leaks, tanks and containers, pumps and valves, compressors, and flanges. These individual sources may then be grouped into source categories and subcategories, as identified by the EPA.

An important factor in regulating air toxics is emission rates. Emission rates can be determined in a variety of ways, including continuous monitoring, source testing, engineering calculations, emission factors, and mass balance.

Source testing provides the most accurate assessment of emission rates. The mass balance approach is as good as the data used in the calculations. A good hazardous materials tracking system provides the best data for a mass balance estimate. If a major source reduces its emissions by 90% (1987 baseline) before MACT is defined, the source may be granted a 6-year extension to the compliance deadline. The EPA published guidance for establishing emissions for synthetic organic chemical manufacturing, chromium electroplating/anodizing, and ethylene oxide sterilization (EPA 1991). Additional guidance will be developed for other source categories. These may be a useful reference document for determining emissions.

Report Preparation

The report should discuss specific areas of noncompliance and concerns regarding poor management practice. It should also include an inventory of emission sources, a description of emission points, an evaluation of emission rates, and identification of existing pollution control equipment.

Develop a strategy for compliance considering federal, state, and local requirements. Many states have adopted programs that go beyond the 1990 CAA Amendments. In California, the South Coast Air Quality Management District restricts emissions of many of the 189 air toxics. The Arizona Department of Environmental Quality regulates about 200 additional toxics, with a lower threshold for major sources.

The strategy should evaluate options for reducing emissions, including source elimination, product substitution, process modification, and end-of-the pipe control technologies. In the field audit, redundant and unnecessary equipment may be identified that can be completely eliminated. More "environmentally friendly" materials may be substituted, such as powder coat technologies for traditional paints, terpenes for solvents, and hydrofluorocarbons (HFC) or hydrochlorofluorocarbons (HCFC) as refrigerants.

To make a product substitution, it is often necessary to modify the process. Developing a strategy for reducing air toxics and ozone-depleting chemicals will require planning for process modifications, such as new compressor technology for HCFCs and HFCs. It may be necessary to modify a vapor degreaser to substitute terpenes; including adding ultrasonic bottom and sides, removing the chiller, adjusting temperature controls and height controls, and adding a rinse tank and air drying (Green 1991).

The report should look to future requirement that will impact the facility and identify resource needs.

Source Reduction Opportunities and Audit Findings

Having reviewed the facility's procurement, distribution, process, and pollution control equipment data, source reduction opportunities can be identified and prioritized. Typical opportunities or recommendations will include:

- Better communications between chemical users, procurement, and the facility's environmental management group to reduce use of chemicals producing toxic emissions;
- The need for employee training to detect problems or improve chemical process applications;
- Identification of faulty or ineffective pollution control equipment;
- Research on potential chemical substitutes for specific processes that may reduce regulatory and reporting burdens;
- The need for additional monitoring of air emissions to identify and prioritize processes that may require modification;
- Evaluation of specific manufacturing procedures to identify cases where the nature and extent of chemical usage may be altered; and
- Development of an air toxics tracking system to assist with TCSRR report preparation and highlight key areas where reduction strategies must be developed.

A facility may prioritize these and other source reduction opportunities, using criteria such as percent reduction in chemical use achieved in near and long term, rate of return on investment, percent reduction in regulatory, and reporting burden. Facilities are likely to find that the audit targeted for source reduction of air toxics will provide benefits not only on the regulatory front, but also in terms of operational costs, liability, worker safety, and community perception.

References

Green, Bill. 1991. An overview of industrial solvent use or is there life after chlorinated solvents? Proceedings of HMC-South '91 Conference and Exhibition, 24–26 April 1991, in Houston, Texas.

EPA. 1991 *Procedures for Establishing Emissions for Early Reduction Compliance Reductions.*

Walther, Eric, et al. 1991. How to develop your toxic emissions inventory: approaches, problems and solutions. Proceedings of the National Research & Development Conference on the Control of Hazardous Materials. 20–22 February 1991, in Anaheim, California.

14

The Facility Environmental Health Audit

Jacqueline Shields

INTRODUCTION

For the past 3,000 years, environmental health audits have been conducted for many purposes. These audits usually were performed to obtain information and data that could be used in planning and managing the economic status or social order of a community. Although they were not identified by name as an "environmental health audit," they indeed could have been nothing less, even using today's standards as a point of identification.

The environmental health audit is precisely what the name implies. It is an enumeration and description of the health of an individual or community as affected by environmental factors of natural or anthropogenic origin. In the past, it has been conducted with reference to a geographical region or community. Today it is conducted in terms of a specific facility and its impacts on the health of an individual or a community. Although both may accomplish similar objectives, they operate from different rationales.

One major difference in rationale is that of the legal mandate. In the past, the environmental health audit was an outcome of a law that was enacted primarily for economic or political purposes. Over the years, it has been a part or an outgrowth of the census, tallage, or some other similar method of enumeration. These, however, were never specifically identified under law by the title "environmental health audit."

The modern environmental health audit may be directly or indirectly mandated by environmental laws. In the first case, the audit is a distinct activity specified by law, while in the second case it is a secondary or even tertiary activity that must be completed before the primary activity can be initiated. An example of a direct mandate is the environmental health audits required under Sections 104 and 105 of the Comprehensive Environmental Response, Compensation, and Liability Act (CERCLA) and Section 4002 of the Resource Conservation and Recovery Act (RCRA). An example of the indirect mandate is the environmental health audit that is needed to complete an Environmental Impact Statement (EIS) required for the issuance of a federal permit granted under the Federal Water Pollution Control Act or the Clean Water Act (CERCLA; RCRA; CWA)

The second major difference is that of liability and litigation. In the past, environmental health audit information and data were evaluated from the perspective of the economic advantage or disadvantage of a project or action, rather than liability and litigation. In today's litigious society, few, if any, corporations or lending institutions would consider buying a facility with an environmental health history of high rates of occupational or environmental morbidity and/or mortality. There also would be a certain reluctance of any company to purchase or construct a new facility in an area with a known history of adverse environmental health impacts. There would be an equal reluctance to purchase property adjacent to residential areas where class action suits have been filed for compensation for alleged adverse health effects caused by living near the property. These considerations of liability and litigation revolve primarily around the facility.

The third difference is nomenclature. By assigning the name "facility environmental health audit," this becomes a real, identifiable part of the overall environmental audit of a facility. This name quickly distinguishes an environmental health audit done for a facility from the public health–type of environmental health assessment. Further confusion over nomenclature has arisen since the 1980s, when the terms "environmental health audit," "environmental health assessment," and "risk assessment" began to be used interchangeably. This is incorrect, since all three are specific processes with different objectives. Quite simply, the assessment process evaluates, while the auditing process enumerates. By definition, the assessment becomes an extension of the audit by taking the information and/or data collected during the auditing process and evaluating it for a specific purpose.

A "risk assessment" is a specific type of assessment described in the "Risk Assessment Guidelines of 1986" as defining the "adverse health consequences of exposure to toxic agents." It includes hazard identification, dose-response assessment, exposure assessment, and risk characterization, or any combination thereof. In this same document, the "exposure assessment" is defined as a process that calls for a quantification of individuals, populations, or ecosystems exposed or that have the potential to be exposed, to a specified dose of a substance for a certain period of time. The environmental health audit both quantifies and qualifies; however, it does not assign causality or risk. Much of the confusion over "audits" and "assessments" has arisen from the legislation itself (EPA 1987, Cohrssen and Covello 1989; Brown, West, and Bishop 1987).

LEGISLATION

Currently, no law has been passed that was designed specifically for environmental audits and, consequently, environmental health audits. Audits of environmental health are addressed as a part of the legislation that is specific for other environmental areas, such as hazardous waste and contingency planning. Audits also can be found as a part of guideline documents, such as the "Risk Assessment Guidelines of 1986" (EPA 1987). It is unlikely that this situation will change dramatically in the near future because it would be difficult, if not impossible, to develop standards for environmental health. According to the philosophy underlying current legislation, environmental health can be protected and maintained by controlling the exposure source. Also, it would be exceedingly difficult to set standards for "good health" and levy penalties for those who fail to comply. There is, however, an increased emphasis on environmental health assessments and auditing in new and amended environmental legislation. When the previous manor environmental laws (Table 14-1) are compared to the new requirements for environmental health assessment and auditing in the Clean Air Act (CAA) Amendments of 1990 (CAA 1990) (Table 14-2), this becomes evident. Environmental health assessments and audits are required by 43 sections of the 9 environmental laws listed in Table 14-1 (CERCLA; RCRA; CWA; CAA 1990; CAA 1977; FIFRA; NEPA; SDWA; TSCA).

An examination of the summary of the provisions for environmental health audits and assessments in the major environmental laws in Table 14-1 shows limited provisions for audits compared to those for assessments and research. Study and research took precedence over auditing in the early environmental legislation, particularly the laws regulating water pollution. As environmental laws evolved, so did the requirements for assessments, particularly risk assessments. The late 1970s and early 1980s saw a substantial increase in the development and prominence of environmental health auditing. Much of this was the result of the need to not only determine the effectiveness of legislation in controlling pollution to protect public health, but to also evaluate the public health benefits derived from the large expenditures required to comply with these laws.

Of the four sections addressing environmental health in the CAA Amendments of 1977, two contain specific provisions for environmental health auditing (CAA 1977). Section 202 calls for a summary of health

TABLE 14-1 Environmental Health Auditing/Assessment Requirements

Act	Section	Specified Activity
CAA Amendments of 1977	Title I	
	Sec. 202(D)(i)	Summary of health effects
	Sec. 313	Reporting of air pollution episodes and emergencies
	Sec. 402	Research
	Sec. 403(c)(1)	Audit of chemical contamination in humans
CERCLA	Sec. 104(i)	Establish and maintain: 1. National disease registry 2. Literature inventory 3. Listing of closed areas 4. Provide medical care and testing in emergencies 5. Survey and screening programs for illness/toxicologic effects
	Sec. 105(8)(A)	Determination of relative risk to human health, welfare, and environment
	Sec. 301(c)(2)	Assessment of effects protocols for assessments
	Sec. 301(e)(3)(C)	Evidentiary burden to prove adverse effects of releases
CWA	Sec. 104(m)(B)	Study of long-term effects
	Sec. 104(n)(1)	Studies of effects
	Sec. 104(n)(2)	Audit of estuaries
	Sec. 104(r)	Research
	Sec. 104(s)	Studies of effects
	Sec. 104(t)	Studies of effects
	Sec. 105(d)	Identify and measure effects
	Sec. 304(a)(1)(A)(C)	Studies on effects
	Sec. 305	Water quality inventory
	Sec. 403(c)(1)	Determine effects and promulgate guidelines
FIFRA	Sec. 20	Research and monitoring
NEPA—Title I	Sec. 102(2)(C)(i)(ii)	Environmental impacts
	Sec. 204(5)	Conduct investigations, studies, surveys, research, and analysis of ecosystems and environmental quality
	Sec. 204(6)	Document and define changes in plants and animals
RCRA	Sec. 4002(c)	Consideration of demography and ecology
	Sec. 7001(f)	Identify hazards to health

TABLE 14-2 ***(continued)***

Act	Section	Specified Activity
	Sec. 8002(f)(3)	Study and analyze dangers to health and the environment
	Sec. 8002(m)	Study and analyze dangers to health and the environment
	Sec. 8002(n)	Study and analyze dangers to health and the environment
	Sec. 8002(o)	Study and analyze dangers to health and the environment
	Sec. 8002(p)	Study and analyze dangers to health and the environment
SDWA	Sec. 1412(e)(2)	Study health effects and contamination levels
	Sec. 1412(e)(3)(A)	Evaluate impacts on susceptible populations
	Sec. 1412(e)(3)(C)	Evaluate synergistic effects
	Sec. 1442(a)(1)(B)	Identify and measure health effects
TSCA	Sec. 4(a)(1)(A)(B)	Risk assessment and testing of chemicals for health effects
	Sec. 4(b)(2)(A)	Testing; standards development
	Sec. 4(e)(1)(A)(iii)(vi)	Consideration of exposed individuals and extent of exposure
	Sec. 5(a)(2)(B)	Consideration of exposure
	Sec. 5(b)(4)(A)(ii)(I)	Submit data to show no unreasonable risk
	Sec. 5(b)(4)(A)(ii)(II)	Show evaluation of no unreasonable risk
	Sec. 6(c)	Publication of effects, benefits, uses, availability, economic risk
	Sec. 8(a)(2)(E)(F)	Report all existing health data, number of exposures, and estimated exposures
	Sec. 8(d)(1)	Listing of health and safety studies
	Sec. 10(c)	Assessment and monitoring research

effects caused by air pollution, and an audit of chemical contamination in humans is required under Section 403. As noted in Table 14-1, the other two sections address research and the reporting of air pollution episodes and emergencies.

The Clean Water Act of 1972 (CWA) does not contain direct provisions for environmental health audits, but in Section 105 it does direct that the effects of water pollution be identified and measured. Calls for studies, research, assessments, and water quality and estuary inventories compose the majority of this section (CWA).

TABLE 14-2 1990 CAA Amendments Environmental Health Auditing/Assessment Requirements

Title	Section	Subpart	Specified Activity
Title III Hazardous Air Pollutants	Sec. 301 Hazardous Air Pollutants	(d)(4)	Consideration of health threshold when establishing standards
		(m)(5)	Report on the assessment of public health and environmental effects of atmospheric deposition to Great Lakes and coastal waters
		(n)(5)	Assessment of public health and environmental effects from H_2S emissions associated with extraction of oil and natural gas resources
		(n)(6)	Assessment of hydrofluoric acid rare events on pH and the environment
		(o)(1)	Request of NAS for a review of EPA carcinogenic risk assessment methodology
		(o)(2)(A)	Study of techniques for estimating and describing carcinogenic potency
		(o)(2)(B)	Study of techniques for estimating exposure to HAPs
		(o)(3)	Evaluate and report on methodology for assessing inheritable genetic mutations, birth defects, and reproductive dysfunctions
		(o)(5)	Reporting on demand information from testing, records, reports, research, or other activities
		(o)(7)	Establishment of carcinogenic risk assessment guidelines
		(r)(1)	Identify hazards from accidental releases and use appropriate risk assessment techniques to design and maintain facility safety
		(r)(4)(A)	Consideration of severity of acute adverse health effects
		(r)(4)(C)	Consideration of potential magnitude of human exposures
		(r)(6)(F)	Chemical Safety Board to conduct studies and research on potential for accidental releases
		(r)(7)(B)(ii)(I)	Hazard assessment of potential effects of accidental releases, including determination of downwind effects and potential exposures to affected populations
		(r)(7)(B)(iii)	Registration of risk management plan
		(r)(8)	Conduct long-term research to develop and determine information on methods and techniques for hazard assessment
	Sec. 303 Risk Assessment and Management Commission	(b)(1)	Consideration of use and limitations of risk assessment
		(b)(2)	Consideration of most appropriate methods for measuring and describing risks
		(b)(3)	Consideration of methods to reflect uncertainties

TABLE 14-2 ***(continued)***

Title	Section	Subpart	Specified Activity
			in measurement and estimation
		(b)(4)	Consideration of risk management policy issues
Title III	Sec. 303	(b)(5)	Development of a consistent risk assessment methodology among federal agencies
	Sec. 304 Chemical Management Process Safety	(c)(2)	Requirement of employers to perform workplace hazard assessments
Title VII Miscellaneous	Sec. 812 Role of Secondary Standards	(a)(1)	Reporting of information on effects on welfare and environment that are caused by pollutants listed in Sec. 109 and other pollutants that may be listed
Title IX Clean Air Research	Sec. 901 Clean Air Research	(d)(1)(A)	Conduct studies, including epidemiology, clinical, laboratory, and field studies to identify and evaluate exposures and effects
		(d)(1)(A)	Conduct studies, including epidemiology, clinical, laboratory, and field studies to identify and evaluate exposures and effects
		(d)(2)	Conduct research to develop methods and techniques to identify and access accidental exposures to air pollutants
		(d)(2)(B)	Prioritize preparations of environmental health assessments
		(d)(2)(C)	Evaluations to be based on reasonably anticipated toxicity to humans and exposure factors
		(d)(2)(C)(i)	"An examination, summary and evaluation of available toxicological and epidemiologic information for the pollution to ascertain the levels of human exposure which pose a significant threat to human health and the associated acute, subacute, and chronic adverse health effects"
		(d)(1)(C)(ii)	Determine gaps in health effects and exposure level information available
		(e)(2)	Evaluation of risks to ecosystems exposed to air pollutants, including characterization of causes and effects and reversibility of chronic and episodic exposures
		(e)(4)	Evaluation of effects of air pollution on water quality, including assessments of short- and long-term effects of acid deposition and other atmospheric pollutants on surface water (including wetlands and estuaries) and groundwater

TABLE 14-2 ***(continued)***

Title	Section	Subpart	Specified Activity
		(e)(5)	Evaluation of effects of air pollution on forests, materials, crops, biological diversity, soils, and other systems
		(j)(3)(B)(ii)	Maintenance, upgrading, and application of acid deposition models for emission interactions and ecosystem responses to acid deposition
		(j)(3)(E)(iii)	Report on status of ecosystems (including forests and surface waters), materials, and visibility affected by acid deposition
		(j)(3)(E)(iv)	Report on cause and effects of acid deposition on changes in surface water quality, forests, and soil conditions

Specific requirements for environmental health audits are found in Sections 104 and 105 of the Comprehensive Environmental Response, Compensation and Liability Act of 1980 (CERCLA). Under Section 104 (g)(i), provisions are made for:

1. The creation of the Agency for Toxic Substances and Disease Registry (ATSDR);
2. The establishment and maintenance of health registers;
3. Inventories of health studies;
4. Inventories of closed contaminated areas;
5. Provision of medical care and sampling in emergency situations; and
6. Survey and screening programs.

The EPA Administrator is mandated to perform the following:

(1) in cooperation with the States, establish and maintain a national registry of serious diseases and illnesses and a national registry of persons exposed to toxic substances;

(2) establish and maintain inventory of literature, research, and studies on the health effects of toxic substances;

(3) in cooperation with the States, and other agencies of the Federal Government, establish and maintain a complete listing of areas closed to the public or otherwise restricted in use because of toxic substance contamination;

(4) in cases of public health emergencies caused or believed to be caused by exposure to toxic substances, provide medical care and testing to exposed individuals, including but not limited to tissue sampling, chromosomal testing, epidemiological studies, or any other assistance appropriate under the circumstances; and

(5) either independently or as part of other health status survey, conduct periodic survey and screening programs to determine relationships between exposure to toxic substances and illness. In cases of public health emergencies, exposed persons shall be eligible for admission to hospitals and other facilities and services operated or provided by the public Health Service.

The extensive volume of data collected as a result of these requirements has been used to develop health registers at national and local levels and disciplines such as epidemiology (CERCLA).

Further in CERCLA, Section 105 (8)(A), it is stated:

> criteria for determining priorities among releases or threatened releases throughout the United States for the purpose of taking remedial action and, to the extent practicable taking into account the potential urgency of such action for the purpose of taking removal action. Criteria and priorities under this paragraph shall be based upon relative risk or danger to *public health or welfare or the environment*, in the judgment of the President, taking into account the extent possible the *population at risk*, the hazard potential of the hazardous substances at such facilities, the potential for contamination of drinking water supplies, the potential for *direct human contact*, the potential for destruction of *sensitive ecosystems*, State preparedness to assume State costs and responsibilities, and other appropriate factors; (CERCLA)

In CERCLA, Section 107(a)(4)(C), under liability provisions, there is an indirect reference to an environmental health audit in the statement "damages for injury to, destruction of, or loss of natural resources, including the reasonable costs of assessing such injury, destruction, or loss resulting from such a release." (CERCLA) In order to ascertain the amount of such damages, an environmental health audit should be conducted. The other alternative, of course, is to use a formula method of determining damage costs.

The RCRA federal guidelines for regional planning for waste management include considerations of population, density, distribution, and projected growth (RCRA, Sec. 4002 (c)(4)). Again, there is the implication that an audit of environmental health should be performed as a part of the acquisition of these data; however, an environmental health audit is not specified. In no part of the RCRA is there mention of the need to ascertain environmental health baseline data for an area or of the impact that a facility would have on the environmental health of the area (RCRA).

The Environmental Auditing Policy Statement issued by the EPA on November 8, 1985, (F.R. 1985) does not address a direct approach for auditing environmental health. Rather, it is written in terms of compliance and enforcement. The auditing of the impacts of a facility on the environmental health of the area are certainly not discussed (EPA 1985).

In one of the first federal environmental laws, the National Environmental Policy Act (NEPA) in 1970, the emphasis was on quantifying and qualifying environmental health effects. NEPA required that changes in plants and animals be documented and that investigations, studies, surveys, research, and analysis be conducted for those ecosystems and the environmental parameters that would be impacted by certain projects or actions. Audits were required in two of the four sections that addressed environmental health (NEPA).

In the CWA of 1972, three of the ten sections pertaining to health provide for environmental health auditing (CWA). Section 104 calls for an audit of estuaries, Section 105 for an identification and measurement of effects of water pollution, and Section 403 for a determination of effects. The requirements for environmental health auditing were expanded in the Safe Drinking Water Act (SDWA) of 1974 to include two provisions for auditing in Section 1412 and one in Section 1442 (SDWA). This is three out of four provisions in the act for health effects assessment and auditing.

Later in 1976, with the passage of the Toxic Substances Control Act (TSCA), the scope of auditing was expanded when a report of all existing health data and a count of all exposures and estimation of exposures was required by Section 8. Section 4 of TSCA also stated that exposed individuals and the extent of the exposure be considered. The other eight provisions for health evaluations included risk assessment, health assessment, testing, evaluations, and reporting of studies and literature searches (TSCA). Again the emphasis was on assessment.

The assessment of environmental health effects has been greatly expanded in the CAA Amendments of 1990, as shown in Table 14-2 (CAA 1990). There are 37 specific provisions for either assessment or auditing in the act. Section 301 of Title III, Hazardous Air Pollutants, contains 17 distinct provisions; however, only three call for auditing. Section 301(o)(5) requires the reporting on demand of information obtained in testing, research, or other activities or from reports or records. This implies a demand for a more thorough examination of existing information, rather than requiring the generation of new data. Further provisions in Section 310(r) will necessitate a thorough examination of the quantity and quality of health effects associated with activities conducted at a facility. The determination of downwind effects on affected populations is mandated in Section 310(r) as well as a consideration of the severity of acute adverse health effects. This would include community health effects as well as impacts on the health of individuals. All of these call for auditing rather than for assessing exposures and potential for exposures that are addressed in the remaining 14 subparts of Section 310 listed in Table 14-2.

The emphasis on risk assessment and management in the 1990 CAA Amendments is highlighted by the assignment of one entire section to these activities. Section 304 not only establishes a Risk Assessment and Management Commission, but it addresses methods for measuring and estimating risk. Risk assessments can be performed using existing health information or data from a facility or from comparable facilities or situations. Given the attendant uncertainties and assumptions in the risk assessment process, more realistic risk values may be obtained using actual baseline data from the facility (EPA 1987; Cohrssen and Covello 1989).

Further auditing of environmental health effects is called for in Section 812 of Title VII. This provides for the reporting of information on the effects on welfare and the environment caused by secondary standard air pollutions.

Finally, in Section 901, Title IX, Clean Air Research, only 3 of the 13 parts related to environmental health call for identifying and reporting on health and welfare effects. Section 901(d)(2)(C)(i) specifically calls for "an examination, summary and evaluation of available toxicologic and epidemiological information for the pollution to ascertain the levels of human exposure which pose a significant threat to human health and the associated, acute, subacute, and chronic adverse health effects." This is a definite expansion of the scope of auditing with the inclusion of subacute adverse health effects. This may present particular difficulties, since these subacute effects may be confused with those produced by exposures from other sources or with the effects of such confounding variables as smoking and alcohol consumption.

Section 901(j)(3)(E)(iii) also expands environmental health auditing to include ecosystems, materials, and visibility affected by acid deposition. This, however, is more in the realm of regional environmental health auditing and less related to the environmental health auditing done for a facility.

Finally, in Section 901(d)(1)(C)(ii), a determination of the deficiencies of environmental health information and data is required (Table 14-2). This may be one of the more important provisions of the entire act because it will greatly expand the scope of research to establish baselines for exposures. The lack of such data and information has led to one of the major problems in health assessments, which is overestimating risk using inappropriate assumptions. In the past, there have been few alternatives to using the standard assumptions, given the dearth of environmental health baseline data and information. Identifying these deficiencies will also stimulate research in those areas where knowledge has been lacking in the basic science and mechanisms of pollution and its effects on health and the environment. While this may appear at first glance to be a costly requirement, it may indeed be more cost-effective in the long term by focusing research on real rather than perceived problems. This also can lead to more efficient and proactive monitoring systems and strategies.

Although the CAA Amendments of 1990 have increased the requirements for environmental health auditing and assessment more than in all previous environmental legislation, this does not need to be regarded as a negative factor. Environmental health auditing is the mechanism whereby environmental health baselines are established. This baseline data and information can then be used to design and operate more efficient monitoring systems, reduce liability by properly identifying exposure sources, and produce more realistic risk assessments. Also, it should be noted that many of these assessments pertain to communities or regions and not to a facility. Prudent administrators and managers will, however, conduct an environmental health audit to establish an environmental health baseline for the facility prior to agency demand. The voluntary performance of an environmental health audit also demonstrates "good faith" and the desire to be a "good neighbor" to other facilities and the surrounding community, thus serving as a bridge in building better community relations. Facility managers and corporate leaders can use the environmental health audit information and data to identify and solve existing problems rather than needlessly expending time, money, and resources in a defense against perceived problems.

STRUCTURE OF THE ENVIRONMENTAL HEALTH AUDIT

Since the environmental health audit has not been specified in legislation or guidelines, there is no standardized audit form. The environmental audit form shown in Figure 14-1 is generic and can be used to audit health effects associated with water, hazardous waste, or any other source of exposure. Data and information can be entered on the form in a variety of ways. One way would be to simply check the appropriate area for which data are available and add a reference number to indicate where data and information can be found. Some will prefer to use only the reference number to denote the availability of data and information. Others will prefer to enter data and information directly on the form. Certainly, it can be modified to meet the requirements and needs of a specific facility or standardized for use by multiple facilities within a particular industry. This discussion of the structure and use of the audit, however, will focus on air pollution exposures. A complete environmental health audit should consist of four major categories: Population, Ecosystem, Exposure, and Treatment. Each category is then audited systematically, as shown in Figure 14-2, with respect to operations at the facility and emergency situations that have occurred at the facility. It should also be remembered that the entire environmental health audit is conducted from the perspective of the facility and not as a public health audit or a community health assessment.

The first category to be considered is that of Population. This includes the total population in the surrounding communities, the individuals at the facility, the affected community, documented clinical manifestations, and

Audit Category	Operational		Emergency	
	Community	Individual	Community	Individual
POPULATION				
Total population				
Affected population				
Clinical manifestations				
Morbidity (incidence, rates)				
Mortality (incidence, rates)				
Psychological impacts				
Documentation				
ECOSYSTEM				
Flora				
Enumeration				
Habitat				
Morbidity				
Mortality				
Ecological niches				
Threatened and endangered species				
Fauna				
Enumeration				
Habitat				
Morbidity				
Mortality				
Ecological niches				
Threatened and endangered species				
EXPOSURE				
Source				
Route				
Media				
Quantity of contaminant				
Quality of contaminant				
Dose				
Duration of exposure (continuous, intermittent)				
Type (additive, synergistic)				
TREATMENT				
Clinical				
Public health				
Removal of exposure				

FIGURE 14-1. The Environmental Health Audit.

Audit Category	Operational		Emergency	
	Community	Individual	Community	Individual
Mitigation of exposure				
Protection of unexposed population				
Recovery				
CONSEQUENCES				
Economic				
Productivity				
Consumption				
Income (loss, gain)				
Costs (health, non–health related)				
Benefits (health, non–health related)				
Resource utilization				
Social				
Community status				
Ecological status				
Quality of life				
Risk assessment				
Risk management program				
Medical response capabilities				
Existing perceptions				
Potential perception developments				
Community response				
Political				
Existing legislation and regulation				
Existing political mechanisms				
Litigation				
Liability				
Judicial response				
Agency response				
Private sector response				
Business response				
Industry response				
Community response				
Observed changes				
DOCUMENTATION				
Data management system(s)				
Existing data bases				
Information sources				

FIGURE 14-1. *(continued)*

Audit Category	Operational		Emergency	
	Community	Individual	Community	Individual
Collection of data and information				
Storage of data and information				
Situation analysis				
Reporting (data, information, analysis, etc.)				

FIGURE 14-1. (*continued*)

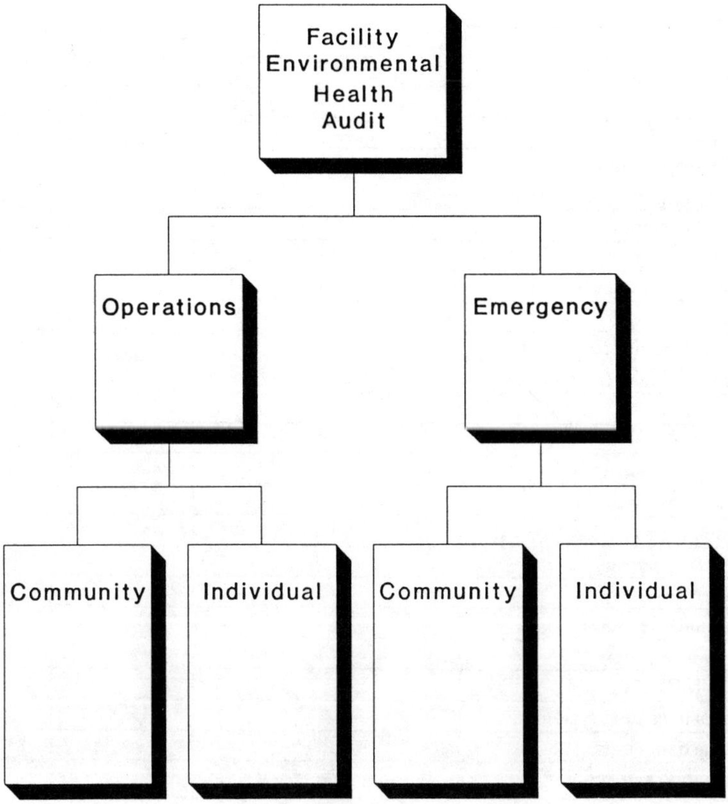

FIGURE 14-2. Systematic auditing of environmental health.

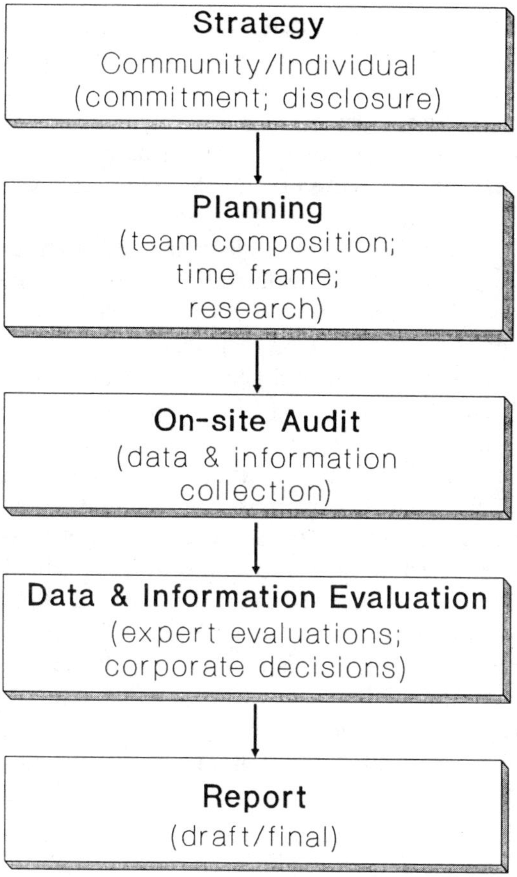

FIGURE 14-3. Progress for facility environmental health audit.

morbidity and morbidity incidence and rates. Anecdotal as well as documented information and data concerning psychological impacts should also be included in the audit since this category is often ignored until litigation threatens. Psychological disturbances experienced by individuals or communities may be the precursors of later clinical manifestations and should not be dismissed. The presence or absence of the documentation used to quantify and qualify all of these categories should also be included. When the documentation is extensive, incorporation by referencing in the audit is recommended. If the data and information are of a sensitive nature, a notation indicating only the presence or absence of documentation may be more appropriate.

The next category is that of the Ecosystem. This is subdivided into Flora and Fauna. Usually, these terms denote wildlife and flora found in natural habitats; however, in the environmental health audit, it has been expanded to include both domestic animals and crops. Usually, the effects of domestic animal or crop exposures are considered only if financial restitution for damages is involved. An audit done in the aftermath of a sudden, unplanned, and uncontrolled event will focus on only those effects that obviously can be attributed to the event. It may include effects that have been wrongly attributed to such an event if a thorough environmental health audit has not been previously performed at the facility in question. This increases the difficulty of evaluating the relationship between changes in environmental health effects and the emissions from a facility.

A comprehensive environmental health audit, as shown in Figure 14-1, will consider health effects resulting from both operational and emergency exposures. Again, the environmental health of domestic animals and crops are audited in terms of their presence in and around the facility and the impacts of the facility on each. A thorough examination of threatened and endangered species (T&Es), habitat, and ecological niches that have been or will potentially be affected by facility operations is particularly important. Unfortunately, these are often excluded in the more traditional health assessments. Any audit that includes historical data and information can be used to determine trends in ecosystem component quantity and/or quality. Historical data can substantiate improvements in ambient air quality in the vicinity of the facility as well as reductions of air emissions from the facility. Depending on the trends observed, the potential for liability may be reduced if favorable trends occur or if mitigation measures can be instituted in the case of negative ones.

Although the Exposure category does not assess health effects, it does identify those sources and substances that have or possess the potential to affect environmental health at the facility or in nearby communities. Particular emphasis should be placed on a careful examination of each of the 189 hazardous air pollutants listed in Title III of the CAA Amendments of 1990 as a source of exposure. These will receive particular attention by regulatory agencies during the course of facility inspections and also in community health assessments. Sources, routes of pollution migration, and affected media are directly related to the quantity and quality of the contaminant. The dose, duration of exposure, and type of exposure provide the baseline information for health assessments. It is important to note if each has been calculated or measured.

Often, doses are estimated or calculated on the basis of estimated quantities of releases. The probability of obtaining measured dosage data is higher for individuals working at the facility, since they may be routinely equipped with personal monitors. Community dosage data is usually estimated, since personal monitors are not commonly worn by the average citizen. The exception would be if a special health study requiring the use of personal monitors for specific substances is being conducted in the area. Even then, the monitored substance may not be the one involved in the exposure incident. It is also important to audit the duration of an exposure in terms of being continuous or intermittent in nature, particularly if subacute or chronic health effects are under consideration. Continuous exposures usually are associated with these types of health effects, and intermittent exposures are associated with acute health effects. Intermittent exposures also tend to be those associated with emergency situations and consequently involve higher quantities released into the environment. These exposures often produce the acute types of environmental health effects that are obvious and easily diagnosed.

The treatment category is a confirmation and documentation of environmental health effects resulting from exposures to emissions from the facility. These have occurred in individuals at the facility and/or in the surrounding communities. It includes both clinical and public health reports of the health effects caused by exposures. The method and time involved in the removing or mitigating the exposure also should be included. Methods used to protect the unexposed population should be noted. Complete records of the recovery of individuals, the community, and operations of the facility following an exposure incident may be particularly difficult to obtain. The recovery of an individual's or community's health to levels existent prior to the exposure may not be possible. In such cases, the audit should include the degree of recovery that can be accomplished.

The fourth category, Consequences, has been subdivided into three subcategories: Economic, Social, and Political. Each provides an index of the impact of environmental health effects on the facility in terms of potential enforcement actions, liability, community relations, and, last but not least, the potential to affect future environmental legislation. To ignore any of these in the environmental health auditing process deprives the owners and management of a facility of data and information for future use in defense of past performance or initiation of proactive economic or environmental programs. Certainly, the information and data can be used not only in future health and risk assessments, but also as guides in planning and management.

The secondary effects of environmental health on the economic, social, and political status of the facility also may be documented in this category. Once they have been identified, it is then possible to qualify and quantify how environmental health has affected the productivity of the facility. The profit or income in terms of loss or gain is directly tied to costs associated with maintaining, restoring, or protecting environmental health. The benefits derived from these expenditures may be short term or long term. The long-term variety can increase over time as a result of maintaining, restoring, or protecting environmental health. An example of this would be the lack of community opposition at a permit hearing for a facility that has demonstrated good faith and community concern in its environmental health expenditures or efforts.

The Social subcategory is actually an environmental health sentinel category. The items included in this subcategory may be the first indicators that environmental health is being affected by or affecting facility operations. This is particularly important when subclinical health effects are involved or are in question. Too often, a blighted area surrounding a facility and an unfriendly nearby community are ignored and undocumented until the agency is knocking at the front gate and the neighbors are marching outside with pickets. Attending to and documenting these, when the first warning signs appear, allow the owners and facility manager to operate from a proactive rather than a defensive posture. Community responses are often based on the perception of how the health of that community has been affected by a nearby facility. While this may or may not be true, it is nevertheless a fact of modern life. It is therefore essential that close monitoring and documentation of community perceptions and responses plus the anticipation of developing perceptions be done if these problems are to be resolved. Failure to do so can only result in costly and unpleasant future confrontations among the facility, community, and regulatory agencies.

Documenting the environmental health assessment and risk assessment is essential to an effective environmental health management plan. This, combined with establishing a risk management program, can serve not only to reduce liability and the potential for emergency situations, but also can be used to develop positive community relations.

The Social category of the environmental health audit should also include information concerning the area capabilities of health care facilities and health care providers to treat any health effects that may be caused by facility operations. This capability is not as prevalent as one would think. Few large cities and almost no small towns have hospitals with emergency rooms, EMS units, or fire departments equipped to respond to incidents involving chemical contaminations of victims. It is prudent, therefore, to determine if a physician or other health care provider capable of treating such cases is present in the area. Even though an occupational medicine physician may be employed at the facility, this may not meet all needs in situations where the environmental health of the nearby community is involved. Working with community leaders to provide resources for environmental health care can be most effective in improving community relations.

Auditing those political areas that affect the facility environmental health status provides not only the basis for prudent management, but also is critical for success in long-range planning for facility operations. To date, environmental health regulation has mainly consisted of requiring that health and risk assessments be done. This is changing as area restoration requirements and financial restitution for health impairment are growing. A thorough review and documentation of the existing legislation regulating environmental health is the first step in assessing any associated litigation or liability that might arise. The existing political mechanisms that can be used to resolve these situations should also be noted. Certainly, the compliance record should be reviewed for any connection with environmental health. If possible, these records should be used to construct a historical profile of improvements and innovations of the facility in environmental health. Information also should be collected concerning the response to the status of environmental health associated with the facility by all affected parties: judicial, agency, private sector, business, industry, and communities. Any observed changes in attitudes or legal requirements should be promptly addressed. The environmental health audit can prove invaluable in monitoring these changes, since it serves not only as a "red flagging" mechanism, but also as a measurement of sensitive changes and to predict trends.

The last category, Documentation, is perhaps the most difficult, not in terms of "being able to do," but in terms of "how to do it." One of the first questions to arise is how will documenting of adverse environmental health effects contribute to liability, litigation, and enforcement actions. Needless to say, it will; however, if these causes do exist, then someone, somewhere will obtain the evidence (i.e., documentation). The most prudent posture, therefore, would appear to be a proactive one. This means initiating an environmental health audit before the health and/or risk assessment is required. This is a corporate decision. If the decision is made to make the environmental health audit an integral part of the complete facility audit, then the documentation process needs to be audited. Disclosure decisions should also be made regarding the protection of information sources.

The environmental health audit data and information can be placed in the same data repository and management system used for the facility audit or in a sole purpose system. Whether one or more data systems are maintained, certain elements will be constant and should also be audited. These include the data management system, methods of data and information storage, collection, analysis, and reporting. Existing data bases that have been used to obtain data and information also should be referenced in the audit.

CONDUCTING THE ENVIRONMENTAL HEALTH AUDIT

As with all successful audits, the full support at the corporate level is required and should be secured before any serious planning is initiated. This is even more important when environmental health is included as a part of the facility audit. Community exposures, real or perceived, are sensitive issues that are difficult to face in the best of circumstances and even more so in the adversarial climate that usually surrounds exposures from facilities. This is exacerbated when a community must, on a daily basis, view facility air emissions that are constant reminders of exposures that potentially can produce the most dreaded environmental health effects, such as cancer or birth defects. These problems and issues often produce a highly charged, emotional climate that is not the best forum for decision making.

There is a growing awareness of the impact of environmental health problems by communities and also by business. Lending institutions are now aware that liabilities can accompany the sale of a facility that has a poor record of compliance and an extensive history of environmental health problems. The increased requirements for health and risk assessments in the CAA Amendments of 1990 will greatly expand the need for the data and information required to provide the baselines for these assessments. The strategies for conducting facility audits should be adjusted accordingly to meet these new requirements. The strategy of conducting health audits should include the following considerations:

- Conducting a complete and impartial audit;
- Determining the current and historical profile of the environmental health status of the facility;
- Using the audit results to identify exposure sources and selecting methods to reduce those exposures;
- Disclosing the audit results; and
- Developing proactive environmental health programs.

While the environmental health audit may or may not be conducted as a part of the facility audit, it must be conducted in the same complete and impartial manner. Such an approach helps establish credibility and a proactive posture early in the environmental health management process. A thorough and unbiased audit can also correct errors in previously assigning exposure liabilities to facility sources.

The audit of the current environmental health status associated with the facility can be accomplished in a relatively short period, since much of the data and information will be readily available. Once this is done, the audit of the environmental health historical profile can be initiated. This, of course, will take longer, since research for existing records and data base inquiries can be a lengthy and time-consuming process. Using a two-tiered auditing process, the current environmental health status can be determined first, with the results being used in the design of the audit of the long-term environmental health profile of the facility.

Much of the disclosure decision, of course, will be made in the same context as the facility audit; however, the sensitivity of the information will also be a large issue in disclosure decisions. The decision for disclosure does not have to be made in a vacuum. A panel composed of medical and scientific experts, legal counsel, the environmental health auditors, corporate executives, and facility management can be convened to assist in the decision-making process. The experience of other facilities with similar disclosure situations can also be used to develop facility environmental health disclosure guidelines.

Options for selecting the team to conduct the environmental health audit include using:

1. The original facility audit team;
2. A specially selected health audit team; or
3. A team composed of members of the facility audit team and medical and scientific personnel.

The first option may be unsatisfactory, since the facility audit team may not possess the skills or body of knowledge required to recognize and effectively evaluate environmental health effects. The use of a team composed of only medical and scientific personnel may also not be a viable option, particularly if they have no knowledge of facility operations. A team composed of members of the facility audit team and medical/scientific personnel would appear to be the option that would produce the most desirable combination of knowledge of health effects and facility operations experience required to conduct an environmental health audit. Because of the diversity of areas that need to be evaluated, the assistance of additional specialist areas may be required. In addition to facility audit team members, other health audit team members may include toxicologists, biologists, zoologists, demographers, medical librarians, health care economists, health care policy analysts, psychologists, nurses, veterinarian, and physicians (e.g., family practitioners, pediatricians, preventive medicine physicians, occupational medicine physicians). These may be employed as full-time team members or on a part-time contract basis.

Acute, visible health effects are usually associated with emergency situations involving sudden, unplanned, uncontrolled releases. Low-level, long-term exposures are more likely to produce chronic, obscure types of environmental health effects. An audit for visible health effects that can be definitively attributed to emission exposures resulting from facility operations is rather straightforward and relatively easy to perform. There may be no visible evidence of environmental health effects related to facility operations; however, sufficient allegations may exist to warrant further investigation. In this case, the first step would be to identify the actual quantity and quality of emissions from the facility. Those emissions then can be classified as National Ambient Air Quality Standards (NAAQS), National Emission Standards for Hazardous Air Pollutants (NESHAPs) under the CAA Amendments of 1977, or any of the 189 hazardous air pollutants (HAPs) listed in the CAA Amendments of 1990. Comparisons of the alleged health effects can then be made with information from existing data bases on the health effects associated with each of these. If the comparison indicates a positive association, further investigations can be conducted. If the comparison, however, indicates a negative association, other exposure sources external to the facility should be considered.

Since many exposures are low level and long term in nature, much of the environmental health audit may consist of searching data bases and information sources rather than physical evidence. There are many types of existing data bases containing health data and information. These may be located at the facility or may exist in the public or private sector.

Public data bases include cancer and birth defects registries maintained by state and federal agencies; records of publicly funded institutions, such as hospitals, nursing homes, fire departments, emergency management offices, and so on; and special health effects registries maintained by ATSDR and the EPA. Health effects information can also be obtained from poison control centers and the National Response Center. Private sector data bases also exist; however, access may be restricted and/or include a user fee. Some form of a user fee is usually required to access computerized health and medical data bases, such as MEDLINE. Others, such as CHEMTREC™, are financed by subscribers' fees, although they can be accessed for free by non-subscribers in emergency situations.

Although environmental health data and information can be found in private sector data bases, these may be difficult to access because of confidentiality restrictions or the adversarial position of the organization. An example of a private sector data base would be health care and life insurance company records. These are good sources of morbidity and mortality data. The company may have already completed much of the work required to audit the environmental health of communities near the facility. Labor union records are another good source; however, it may be difficult to access these data bases. Other private sector data bases include those produced by citizen activists. Many citizen activist groups are now using personal computers to establish environmental health data bases for specifically affected areas. These should not be overlooked when conducting the environmental health audit of a facility, particularly when it is located in one of these affected areas.

When the audit has been completed, the final evaluation of the results should be conducted by a team of medical and scientific experts, the facility manager and designated facility personnel, the health audit team, and representatives from the facility audit team. Using a neutral facilitator to conduct the health audit evaluation discussions is recommended. This will assist in resolving controversies that may arise when interpreting results of the audit, and speed the evaluation process.

SUMMARY

The benefits of conducting an environmental health audit far outweigh the disadvantages. The immediate advantage is to place the facility in a position to comply with existing health and risk assessment requirements, particularly those contained in the CAA Amendments of 1990. Benefits can be accrued by correctly identifying exposure sources, particularly those not related to emissions from facility operations. A thorough audit can be used to dispel allegations of facility emissions effects on environmental health, thus reducing attendant liability. The audit also can be used to identify sources of potential exposure and enable the facility manager to take corrective measures, thus minimizing future liability. As with the environmental audit of the facility, there is the concern that the results of the health audit will be used to penalize the facility for past activities and exposures. While this is a real concern, the larger question remains of the litigation and liability associated with unidentified air emissions that continue to affect environmental health.

References

Comprehensive Environmental Response, Compensation, and Liability Act, PL 96-510, 42 U.S.C. 9601.

Resource Conservation and Recovery Act, PL94-580, 42 U.S.C. 6901.

Clean Water Act, PL 92-500, 33 U.S.C. 1251.

Environmental Protection Agency. 1987. *The Risk Assessment Guidelines of 1986.* EPA/600/8-87/045, U.S. EPA, Washington, D.C.

Cohrssen, J. J., and J. T. Covello. 1989. *Risk Analysis: A Guide to Principles and Methods for Analyzing Health and Environmental Risks*, Council on Environmental Quality, Washington, D.C., 1989.

Brown, H. S., C. R. West, and D. R. Bishop. 1987. Chemical health effects assessment methodology for airborne contaminants. *Risk Analysis*, Vol. 7, No. 3, pp. 389–402.

Clean Air Act Amendments of 1990, PL 101-549, November 15, 1990.

Clean Air Act Amendments of 1977, PL 95-95, 42 U.S.C. 1957.

Federal Insecticide, Fungicide and Rodenticide Act, PL 94-140, 7 U.S.C. 136.

National Environmental Policy Act, PL 91-190, 42 U.S.C. 4321.

Safe Drinking Water Act Amendments of 1977, PL 95-190, 42 U.S.C. 300F.

Toxic Substances Control Act, PL 94-469, 25 U.S.C. 2601.

Environmental Protection Agency. Environmental auditing policy statement. *Federal Register*, Vol. 50, No. 217, November 8, 1985, pp. 46504–46508.

Shields, J. 1990. *Environmental Health: New Directions*, Princeton, N.J.: Princeton Scientific Publishing Co.

IV

Case Studies

Case Study #1

Reductions in Air Emissions of Ozone-Depleting Chemicals

Pamela J. Beilke

As a result of the requirement to submit a toxic release report under Section 313 of the Superfund Amendments and Reauthorization Act (SARA), an electronics manufacturing facility realized that over 50% of the solvents used were released into the air. The primary solvents in use at the time were freon and trichloroethane (TCA), both ozone-depleting chemicals (ODCs). The facility made a commitment to reduce the air emissions of these ODCs and ultimately eliminate their use.

An auditing approach was used to identify the sources of ODC air emissions. The preaudit included a review of the following information to identify where and how freon and TCA were used:

- Material safety data sheets—To identify chemical products which contained freon and/or TCA.
- Chemical inventory—To identify locations where the chemical products were used.
- Procurement records—To determine the quantities purchased.
- Air permits—To identify processes and equipment that utilize freon and TCA.
- Floor plans—To verify the locations of processes and equipment.
- Production records—To evaluate the utilization of equipment.

Much of this data had already been collected during the preparation of the SARA 313 report.

During the field audit, permitted equipment was inspected and intermittent and non-process activities were evaluated. The objective was to verify proper operation and maintenance and identify opportunities for reducing emissions. The options considered included:

- Source elimination;
- Product substitution;
- Process modification; and
- Control technology.

The assessment identified processes where alternate materials could be substituted for the ODCs. In particular, spray cans of Freon™ were used for general cleaning purposes. For many manual cleaning uses, isopropyl alcohol was a suitable substitute.

The review of production and equipment usage records revealed that many degreasers were underutilized. Very often, two or more degreasers were in close proximity to one another. Many of the existing degreasers were not being properly operated or maintained. Chilling coils were not adjusted to the proper temperature. Many degreasers did not have covers. Often, the existing covers were not used.

A review of the preventative maintenance program revealed that the solvent was drained from the degreasers on a periodic basis, once a week or once a month, rather than on the basis of solvent quality. Sometimes the degreaser had not been used in between solvent changes.

Following the assessment, a plan of action was prepared. The plan included short-term and long-term action items.

Actions to reduce emissions included:

- Consolidate usage and eliminate a number of degreasing units.
- Adjust equipment to optimize operations and minimize emissions.
- Provide training and discipline to ensure proper operations, including the use of covers.
- Install covers where needed.
- Drain and change degreasers when solvent is no longer effective.

Action items to eliminate usage of ODCs included:

- Eliminate cleaning steps when feasible.
- Substitute more acceptable cleaning products.
- Budget for purchase of new aqueous cleaning systems to replace degreasers.

As a result of these efforts, the facility usage and air emissions of freon and TCA were reduced by more than 50%.

Case Study #2

Phased Chemical Usage Study and Authorization Review—Electronics Manufacturer

Linda Micale

A phased chemical usage study was performed over the course of several years for a manufacturing facility whose operations were regulated under a number of environmental laws, including the Emergency Planning and Community Right-to-Know-Act, the Clean Air Act, and OSHA requirements, among others.

First a review of Material Safety Data Sheets (MSDS) was conducted. Then other records on over 3,000 products were reviewed to compile data on the composition, quantity, storage, and use of listed chemical products. When available data appeared to be insufficient or inaccurate, facility managers were interviewed, chemical manufacturers were contacted, and chemical products used were inventoried. The purposes of these activities were to identify authorized products not actually being used, update changes in product information, and develop accurate computerized data bases for various environmental reporting and management purposes.

A later phase of the same project involved a complete facility inventory of chemicals in stock in all chemical-using departments as well as identification of potential or authorized discharged or emission points within each. This information was complied into a computerized data base.

Computerized data bases improved reporting efficiency by allowing a facility to track and update relevant information. Reporting data included quantities of chemicals used, disposed, and emitted; location of storage and use; and associated chemical and physical hazards. A computer program was developed to prepare reports by:

- Comparing chemical constituents present on site to chemicals on various regulatory lists;
- Eliminating constituents in exempt departments or in departments with usage below regulatory thresholds;
- Calculating quantities on site and comparing them to reportable quantities;
- Estimating releases, waste generated, or emissions; and
- Identifying storage and location codes.

In addition to assisting with reporting needs, this project evolved into a centralized chemical authorization and review program. The centralized chemical review process, based upon compiled data and earlier audits, enabled the facility to address concerns related to:

- Preventing employee exposure to toxic chemicals;
- Maintaining chemical hazard information;
- Identifying substitutes for hazardous or toxic chemicals; and
- Minimizing emissions or other wastes produced.

The authorization program involved the continual evaluation of new and updated MSDS, requests for specific chemicals, and user inventories of chemicals within a department. Chemical and physical hazards related to use, storage, equipment, and waste disposal were evaluated.

As in the reporting component of this project, the data base used for chemical tracking is a strong management tool for hazardous materials. The data base contains key information from the department, allowing the facility to track which chemicals are used, as well as where and how much of the chemical has been authorized for use. The review process prompts both the requesting department and the authorization risk management team to consider substitutes or process modifications for particularly toxic substances. The information on the types of waste produced permits identification of opportunities for reuse of recycling. Information entered into the data base must be carefully checked for accuracy and completeness. Quality control checks can include such factors as consistency among different departments for product disposal, waste characterization, and safety requirements.

Case Study #3

Hazardous Materials System and Waste Minimization Project—Electronics Manufacturing Facility

Linda Micale

In 1990, a national electronics manufacturing company conducted a working study to define the basic requirements for an ongoing waste minimization program. The company's approach emphasized the elimination of materials or wastes perceived to have environmental, worker safety, or public health concerns plus evaluating high volumes and costs. In addition to managing a variety of hazardous and industrial wastes, the company reported significant fugitive and point source emissions on the Form R report required each year under the Emergency Planning and Community Right-to-Know Act. Three major components of this working study included:

- Provide a hazardous materials definition to guide the development of waste minimization opportunities and a materials control and tracking system.
- Develop potential minimization opportunities through production area audits and identify "preferred materials" in concert with company engineering and environmental personnel.
- Develop interface requirements for the company's hazardous materials control and tracking system.

The first component—defining hazardous materials—was essential to establish a common management target for the diverse groups involved in such activities at the company. These groups included staff members in purchasing, risk management, chemical distribution, waste management research and development, and operations. The company was convinced that identifying and implementing waste minimization opportunities required the input and coordination of each group. A basic understanding of target chemicals was the first step in the program's long-term success.

The second component comprised a substantial portion of the project. This component involved auditing selected process areas to identify chemical usage, inflow and outflow, processes in place, and chemical/process handling procedures. Also, requests for hazardous material purchases through the procurement department were evaluated, and the stock and distribution areas were toured. Data from chemical risk management software were merged to the extent feasible with audit compilations and fed into an algorithm that ranked the environmental risk of using certain chemicals based on a number of specific factors. The factors can be expanded or revised as needed to match the concerns of the facility. The results of the algorithm provide feedback to facility environmental managers, to design and process engineers and other interested staff on such issues as:

- Chemical substitutions that may result in lower toxicity or quantity emissions, potential exposures, and wastes;
- Where off-specification or excess materials could be utilized by another department with a similar need to reduce overall chemical purchases; and
- Where training or procedural changes may be required to result in more effective use of a chemical to reduce waste generation.

These activities were also designed to allow auditors to view a broader picture of chemical utilization and flow through the facility. Although the audits were not performed strictly from an air quality perspective, high-usage chemicals often include those that are regulated or of concern due to associated emissions. Fifty-seven high-volume materials were identified for source reduction purposes through this study.

Case Study #4

Volatile Organic Chemical Inventory—Nuclear Generating Station (NGS)

Linda Micale

Nuclear generating stations (NGSs) were required to submit to the local regulatory agency a summary of amounts of volatile organic compounds (VOCs) that are emitted from the plant annually. The inventory to identify products containing VOCs was initiated through a review of the distribution data base provided by a central warehouse. As in many facilities, not all chemical purchases occurred through a central department. Attempts to fill this data gap were made by contracting individual contractors that work at the plant and by requesting information on their on-site chemical usage or storage. Approximately 160 products that potentially emitted VOCs were identified. Next, VOC content was verified through a review of the Material Safety Data Sheets (MSDS). Product manufacturers were contacted when the MSDS lacked adequate information. The quantity of VOCs emitted was then calculated by multiplying issue quantities by percent content and volatility.

The information obtained through these reviews was entered into a computer data base developed to facilitate annual lair emissions reporting. Products were segregated according to primary usage. The program was designed to calculate emission quantities based on issue quantities so that future emission reports could be generated simply by entering usage quantities. Labor saving due to this program was estimated at approximately 200 labor hours.

This inventory met basic reporting requirements. However, should more specific or accurate information be required, facility personnel must perform detailed mass balances or other engineering calculations using emission factors, control equipment specifications, disposal records, and other relevant data.

Case Study #5

Waste Minimization

Paul R. Rosenstock and Sam N. Popowcer

The Chemicals-Are-Us Manufacturing Company makes specialty chemicals for a variety of industrial applications. Their plant consists of three jacket-heated, internal coil–cooled kettles and auxiliary weigh and feed tanks. All three reactors have condensers that use water from a nearby lake for cooling. They also have a number of storage tanks for raw materials and products. The plant started operation in 1947 and had no abatement equipment on any of its kettles or tanks; these items were in place before air pollution laws were enacted and, thus, were exempted from permitting requirements.

The plant had no regulatory problems until around 1986, when a new housing project was built within 100 feet of the fence line. Prior to that, the plant was at least one-half mile from its nearest residential neighbors. As soon as the apartment house was occupied, the plant began to receive numerous odor and noise complaints from the new residents and was soon visited by the state environmental agency inspectors. They received citations for odor and waste management violations as the result of a state inspection. The noise complaints proved to be unjustified.

The plant manager and plant technical people began to look at their operation in the realization that they must make significant modifications to continue in operation. They concluded that the major odor was probably caused by four chemicals: butyl acetate, amyl acetate, isophorone, and butyl mercaptan. All were common raw materials in their processes. The waste violations were caused by storage of a large number of "empty" solvent drums in an outdoor storage area. Some drums were outside so long that they rusted through and were leaking.

As an interim measure, the plant stopped producing the products that they felt caused odor problems. A consultant was hired to measure air emissions from the process vessels during reactions that they considered represented the least and most emissions. These tests showed that the condensers were not efficient and that the emissions were considerably higher than they thought. Modeling showed that there was a potential for exceeding recognized maximum ground-level concentrations. Conversion of the condensers to refrigerated water was helpful in reducing emissions and potential ground-level problems, but did not guarantee that odors would be eliminated, especially in butyl mercaptan–containing products.

Extensive laboratory work showed that the use of ester solvents could be minimized and that decyl mercaptan could be substituted for butyl mercaptan. However, these process changes did not eliminate the possibility of fence line odors. At this point, it was realized that air abatement devices would have to be installed. Scrubbers, charcoal adsorption, cryogenic condensers, and vent incinerators were investigated as potential candidates. Scrubbers were ruled out because the local POTW was at capacity; charcoal was rejected because of poor experiences at a sister plant; and cryogenic condensers were ruled out because many of the materials would freeze at low temperatures.

Vent incineration with heat recovery was selected and eventually installed. Originally, only the kettle vents were specified for incineration, but sampling data indicated that the feed tanks and a walk-in hood where drummed raw materials were weighed had to be connected to the incinerator. A change in drumming temperature and the use of dip pipes for drum filling significantly reduced the emissions from the drumming operation.

The problem of empty drum storage was partially solved by ordering and storing most solvents in bulk. In addition, the plant made arrangements with some suppliers to take back the empty drums, for a fee in some cases. All other empty drums were staged in the process area and taken to a drum cleaning and reclaiming operation on a weekly basis. Finally, closed loop transfer systems were installed on the tanks that contained the most volatile and most odoriferous materials. These changes required air quality permits for the incinerator and closed loop transfer systems.

The situation described above took over 3 years to resolve and involved a consent order with the state agency as part of the settlement resulting from the citations. Legal fees exceeded the fines paid. During these 2 years, the plant did a thorough analysis of all of its operations and was able to reduce waste production by over one-half by process and raw material changes. A significant part of the program was done in house with engineering and stack sampling help from a consultant.

Case Study #6

Refinery Environmental Audit

Paul R. Rosenstock and Sam N. Popowcer

The case history presented herein is a composite of a number of audits and waste minimization projects that were done by the authors. It is obvious that this history cannot cover every eventuality that you will encounter. Rather, the case is structured to present some of the typical situations that you may be confronted with and to offer some solutions to the problems that have already been studied and solved.

A major chemical plant that manufactures both a variety of organic and inorganic commodity chemicals conducted an environmental audit of its operators. The facility has an on-site industrial wastewater treatment plant and generates all of its steam needs. Its immediate problem was fence line odors caused by emissions of several chemicals with low-odor thresholds. In addition, the facility found itself faced with diminishing options for waste disposal due to the shutdown of local landfills and drum reclaimers.

SARA 313 reporting was used as a guide to set up an air testing program. An air emissions inventory was done in which process and building vents as well as major storage tanks were sampled. Fugitive emissions were measured on selected processes, and ambient air monitoring was done around the fence line of the plant. Concurrent with the sampling program, process flow sheets and material balances based on raw material usage were developed. The site permits were inventoried to check for compliance. For air emissions, the measured emissions data were used to model ground-level concentrations in the surrounding community for key organic compounds. This modeling showed that the emissions were well below guideline values for community exposure. A similar approach was taken with analysis of the treated wastewater. Modeling showed that the effluent from the wastewater treatment plant caused no adverse impacts on municipal water intakes downstream of the plant discharge.

The audit made recommendations for modifications that could be made to minimize emissions (wastes). These modifications were reviewed based on cost, ease of implementation, and effect on product quality, and a master compliance plan was developed. Research assistance was required to modify some processes and to make certain that new processes were developed with waste minimization in mind. Some typical modifications were as follows:

- A process that used sulfur dioxide as a Lewis acid catalyst had significant sulfur dioxide emissions. A solid Lewis acid was substituted with no difficulty.
- Many volatile drummed raw materials were weighed into buckets from openhead drums. Closed bung drums with drum pumps were substituted, and small "weigh boxes" were installed in the systems.
- Steam jets were replaced with liquid seal pumps.
- The powei house fuel was changed from Number 6 fuel with 1.0% sulfur to natural gas.
- Substitution of less volatile raw materials and cleaning solvents, where possible.
- A process that involved a solvent transfer step that generated a large volume of waste solvent was converted to a solvent-less process.
- Concentrated wastes were collected at the processes and were prevented from entering the wastewater system. This significantly reduced the BOD loading and concomitant sludge generation.
- Filtration technology was changed to minimize both waste products.
- Where practical, dripless valves, "canned" pumps, and/or longer pipe runs (fewer flanges) were installed.
- Closed circuit loading and unloading was installed for many volatile materials.
- Transfer lines from trucks and rail cars were blown back with nitrogen to clean the lines before they were disconnected.
- Operators were trained to clean up spills immediately and to quickly report spills too large for them to handle.
- All personnel were trained to report all spills and leaks immediately.
- Product transfer lines and storage tank allocations were studied and modified, as required, to minimize cleaning requirements.
- Environmental, health, and safety reviews of all new and modified processes and equipment addressed waste minimization. The importance of this step was obvious, since it trained everyone to think of waste streams as part of the process.

- Operating procedures were modified to address waste generation issues.

In some cases, new process equipment was installed, but this was normally done for economic justifications or capacity. However, waste reductions were usually realized because the new equipment was designed with waste minimization in mind. In addition, there were some processes that could not be modified or that could not bear the cost of modification. These were shut down or transferred to another plant.

The conclusion from this work was that waste minimization pays for itself. In addition, it is a never-ending process that requires constant attention to be successful. It must become part of the "culture" of the operation to be truly successful.

To recapitulate, we have presented some of the key issues confronting the chemical industry and chemical users. For the foreseeable future, waste minimization, source reduction, and even zero discharge will be the main thrust of environmental laws and regulations. To survive in this environment, you will have to become more knowledgeable about your processes and about the potential for modifications to achieve the mandated regulatory goals. In many cases, innovative solutions will have to be implemented, including new, as-yet-to-be developed technologies. Knowledge of your process is the key to establishing a successful waste minimization program, and auditing is the vehicle by which you will measure your success.

Case Study #7

Refinery Environmental Audit

Jill Barson Gilbert

This case study is a companion to Chapter 12, "Environmental Audits of Refineries and Petrochemical Facilities." It contains selected air, water, and waste issues identified during a pre-acquisition environmental audit at a refinery. The audit focused on management systems and was not intended as a rigorous compliance audit.

SITE HISTORY

The refinery is over 50 years old and has expanded several times. The facility encompasses about one thousand acres and processes fuels, lubricating oils, and asphalt. The company operates a chemical plant adjacent to the refinery, which shares some of the utilities, storage tanks, and process units.

ENVIRONMENTAL ISSUES

Solid and Hazardous Waste Management

The refinery is a hazardous waste treatment, storage, and disposal facility under the Resource Conservation and Recovery Act (RCRA) and has state and federal registrations. The primary wastes generated at the plant are API oil/water separator sludges, hazardous and non-hazardous tank bottoms, heat exchanger bundle cleaning sludge and contaminated soils. The site also generates various nonhazardous wastes. Wastes are disposed of both on site and off site; on-site waste disposal includes landfills and landfarms. The past and ongoing on-site landfill and landfarming of hazardous and nonhazardous refinery wastes and asbestos creates potential liabilities.

The RCRA Land Disposal Restriction regulations affect many of the facility's hazardous wastes. If the refinery can no longer apply its hazardous wastes at its landfarm because of the RCRA Land Disposal Restrictions, off-site disposal costs will increase by $350,000 to $2,000,000 annually.

At the time of the audit, the facility had applied for, but not yet received, a RCRA permit. The permitting process requires the company to conduct a Land Treatment Demonstration at the landfarm and a RCRA Facility Investigation at various (non-RCRA) solid waste management units. Surface impoundments and landfills comprise the solid waste management units. The RCRA Facility Investigation likely will require remediation or closure of most of the solid waste management units at a cost of over $100 million.

The refinery has an extensive groundwater monitoring system for its existing land disposal facilities, irrespective of the RCRA Facility Investigation.

The refinery is involved in a consent agreement that addresses two on-site waste disposal facilities. The consent agreement contains a civil penalty and requires meeting a compliance schedule for certain field investigations and monitoring activities. Refinery management indicated that the facility has not been identified as a potentially responsible party at any Superfund sites.

Water Management

The refinery and asphalt complex have separate wastewater treatment systems and discharge points. A single NPDES permit covers all of the process water and stormwater discharges. The refinery has earthen ditches that comprise much of the wastewater collection system. These ditches show significant staining. The wastewater basins are, for the most part, earthen. The unlined wastewater collection system and unlined ponds create potential liability for environmental contamination and large future expenditures for closure and remediation. If the wastewater and/or the sludges in the wastewater impoundments are classified as hazardous under new RCRA regulations, the refinery could incur significant costs to close, remediate, and replace the treatment system.

In the 3 years before the audit, the refinery experienced few excursions above its wastewater discharge permit limits. Earlier, an upset at the adjacent chemical plant impacted the wastewater system at the refinery, indicating possible deficiencies in operating procedures.

Air Emissions Management

The refinery has over 75 separate air emissions permits for sources, ranging from storage tanks to sulfur recovery units. Much of the older equipment is "grandfathered," or exempt from permitting requirements. The "grandfathered" equipment and process units represent about 50% of the refinery's air emissions. New permitting requirements could trigger additional controls and would require a major effort to consolidate permits.

The Air Quality Control Region in which the refinery is located is classified as "non-attainment" for ozone. This could affect the level of air emissions controls required for planned refinery projects, particularly if the equipment emits volatile organic compounds.

Future changes in product specifications will increase operating costs. The regulations concern the volatility of gasoline in the summer months and limits on the amount of sulfur in diesel fuel. In the 3 years before the audit, the refinery had numerous air emissions upsets and process unit shutdowns, but received few citations.

Case Study #8

Purchase of Fertilizer Plant

Jill Barson Gilbert

This case study is a companion to Chapter 11, "Property Transactions." The facts described below show the types of data collected during a Phase 1 Risk Assessment at a fertilizer manufacturing plant.

SITE HISTORY

- 1965: Construction of ammonia plant.
- 1966: Construction of anhydrous ammonia terminal.
- 1970: Sale of assets.
- 1972: Construction of liquefied carbon plant.
- 1973: Construction of urea plant.
- 1975: Merger.
- 1978: Joint venture/additional product lines.

ENVIRONMENTAL ISSUES

Underground Storage Tanks

The company operates seven USTs that store lube oil and chemicals. The tanks range in size from 200 to 8,000 gallons. The tanks have not been tested for integrity and do not have cathodic protection. The USTs are registered with the state.

Active or Inactive Waste Disposal Sites

The site has two closed landfills and one small impoundment that reportedly were used for on-site waste disposal in prior years. Of these three facilities, only one of the landfills has been evaluated and was found to have accepted solid wastes consisting of construction debris and other inert material. This landfill, closed in 1972, is partially on the company property and partially on the property of the local natural gas company. A limited subsurface investigation was conducted by the gas company, but no report was submitted to the chemical company.

Other on-site waste disposal areas consist of a one-quarter acre depression used to dispose of equipment-cleaning wastes and other paint wastes, such as acids, potassium carbonate solutions and trash, and a small irrigation pond that held water with about 500 parts per million (ppm) ammonia. Both areas are inactive, and no additional information exists regarding potential amounts of waste or migration of contaminants.

Soil and Groundwater Contamination

The water, wastewater, and solid waste management practices can serve as an indicator of the potential for future soil and groundwater contamination. Three deep wells supply site water. The drinking water was tested and passes city requirements. The site has a history of frequently exceeding permit limitations of ammonia and chromium levels. A review of state files summarizes regulatory activity with respect to NPDES discharge inspections. The most recent sampling by the state agency indicated that the discharge appears to be within the permit limits.

The process wastewater treatment consists of ammonia stripping, followed by oil/water separation and discharge into an unlined settling pond before final discharge to the Mississippi River. The company has no sampling data to characterize the solids, which may contain chromium. The company manages the solids as nonhazardous waste.

The bulk chemical storage tank farm is diked, but not paved. The rail loading and unloading area is partially paved. A 6,000-gallon aboveground No. 6 fuel oil tank contains tank bottoms from prior use of No. 6 fuel oil to fire heaters. The rail unloading area for No. 6 fuel oil contains an underground sump. The sump is full of waste oil, with pools standing around the structure.

Asbestos

The facility could have asbestos insulation, based on the date of construction. Removed insulation currently is not handled as asbestos. No survey or testing has occurred to check for asbestos.

Polychlorinated Biphenyls (PCBs)

Management stated that all PCBs have been removed from on-site transformers and disposed of in approved facilities. The local utility company confirms management's statement.

Residual Materials and Wastes

The company generates hazardous waste under the appropriate Environmental Protection Agency registrations. Hazardous wastes generated at the site consist of spent catalysts, waste oil, and spent solvents. Both hazardous and nonhazardous wastes are accumulated and stored randomly around the site, with little means of secondary containment. Since the site remains in operation, residual materials and wastes are not an issue, as they may be if the site were to be closed.

Compliance and Legal Issues

The company applied for an NPDES (wastewater discharge) permit renewal. Since the company has frequent chromium and ammonia excursions, they may have difficulty in meeting more stringent limits, if imposed when the permit is renewed. The state has expressed minor concern regarding potential contamination due to the unlined settling pond. The state issued a consent order in 1982 due to excessive levels of nitrogen oxide from the nitric acid plant, and citizens filed several complaints. The plant has been in compliance on a routine basis for the last few years.

CONCERNS

The audit team uncovered the following key concerns found during the Phase 1 Risk Assessment, which could translate into liabilities for the seller or buyer, if not properly managed:

- Untested USTs;
- Uncharacterized inactive waste disposal sites;
- Possible asbestos insulation;
- Waste management practices; and
- Unresolved air and wastewater permit issues.

Case Study #9

A Regional Ambient Air Quality Audit

Jacqueline Shields

Rapid industrial growth and economic development and reports of increased adverse human health effects led to public concern about changes in air quality in a large region in the southwestern part of the United States. Since this region forms part of the international border with Mexico, questions were also being asked about what effects unregulated emissions from facilities located across the border in Mexico were having on air quality in the United States. In an effort to begin to address this question, a study was designed to determine what air quality changes, if any, had occurred in the region. The study design work was suspended when it rapidly became apparent that no audit had ever been done to determine the full extent of existing air quality data and information for the region.

Data and information sources used in the audit consisted of the state air quality control agency, the environmental departments of local health agencies, and personal libraries of environmental professionals living in the region. Copies of listings for the computerized data bases were also obtained. The air quality control programs and the collection of data and information prior to 1970 was largely a function of state and local agencies. The monitoring data for this period were not computerized and were found mainly in old reports and in records of the local agencies and state health department.

After the formation of the Environmental Protection Agency (EPA) in 1970, air quality monitoring and control programs became a shared activity among federal, state, and local agencies. From the 1970s to the 1990s, monitoring and analytical methods changed and quality control programs were initiated. The monitoring data were reported to the EPA as required by law, and occasionally the data would be used for special studies. Only annual summary reports of air quality in the region and special study reports were issued by the EPA and the state regulatory agency.

Work on the audit began by submitting requests to local, state, and federal agencies for data and information access. Lists of computerized data bases were also requested. The lists and copies of old reports were slowly assembled over a 6-month period. Personal contact with agencies and environmental professionals in the region also were made to secure additional data and information that had not been included in the early official reports.

The audit revealed that monitoring had been done in multiple sites in 16 cities and counties in the part of the region within the United States and in 5 cities in Mexico. Although for 35 years ambient air data and information had been collected for the region, the majority existed for only about 17 years. Also, the sampling periods and sites varied within each city and county, and few sites had sampling data in excess of 10 years. During this period, monitoring was done for 166 substances. Of these, 149 were monitored using categorical sampling techniques. For example, one high-volume filter sample taken in the field was returned to the laboratory, where analytical tests were done for 25 metals and 9 other elements. One field grab-sample for pesticides was analyzed for 9 pesticide compounds, one bubbler for 5 gaseous compounds, and one field grab-sample for 100 hydrocarbon compounds. Other monitoring was done for PM-10, benzene soluble organics (BSO), sulfates, nitrates, acid vain, visibility, and speciation by electron microscopy. Continuous air monitoring was done for carbon monoxide, nonmethane hydrocarbons, nitrogen dioxide, ozone, and sulfur dioxide.

Monitoring for TSPs was conducted at 72 sites in all cities and counties. Other monitoring included PM-10 (4 cities); BSO (3 cities); sulfates and nitrates (14 cities and counties); bubblers (3 cities); effects package (9 cities and counties); acid rain (1 city); visibility (2 cities); pesticides (3 cities); hydrocarbons (1 city); and continuous air monitoring for gaseous substances (1 city).

The largest volume of data was collected over the longest period of time in the largest city in the United States portion of the region. Monitoring was conducted for all listed substances and categories, with the exception of pesticides. This was the only city in the region in which continuous monitoring for gaseous substances was done. It was also the only city where visibility, acid rain, and hydrocarbon special studies were conducted.

The data were limited to a 10-year period between 1974 and 1984, in 5 cities on the Mexican side of the border in the region. Most monitoring was not consistent throughout the period and existed for only a few years at each site. Monitoring was done using Hi-volume samplers and analytical testing was done for lead (Pb) and total suspended particulates (TSP). Instrument calibrations and analytical testing was done by agencies in the United States.

Although the initial audit was conducted to inventory the existing ambient air quality data and information for the region, it has also led to many other projects. First the results of the audit yielded data and information to develop and establish ambient air quality baselines for the region. Work is currently underway to construct baselines for seasonal, temporal, and spatial levels of ambient air pollutants. This has led to the design of studies to examine trends in air quality for each city and county and to identify significant differences in air quality among cities and various sites within each city and county, as well as between cities in the United States and Mexico. Missing data are also being examined to determine monitoring needs in each part of the region. More importantly, it has provided substantial information that can be used to expand air quality control programs in the region.

Contributors

Pamela J. Beilke

Pamela J. Beilke manages safety, health, and environmental programs for an aerospace company. In this capacity, she directs the development of programs to ensure compliance with environmental and safety laws, regulations, policies, and practices, and provides implementation guidance to manufacturing operations. She has also directed the development of an industry self-evaluation program to verify compliance with these requirements.

Pamela has been involved in environmental issues for more than 12 years. She spent 5 years with the state of Arizona's environmental regulatory agency, where her responsibilities included water and wastewater compliance and enforcement, water quality management planning, and management of groundwater contamination projects. Pamela then served as an environmental consultant with Dames & Moore, where she directed Environmental Regulatory Services, assisting industry in complying with regulatory requirements and conducting environmental assessments.

She graduated magna cum laude with a Bachelor's degree in Environmental Studies from the Intercollege Program at the University of Minnesota and completed additional graduate coursework in Environmental Engineering.

David A. Dougall

David A. Dougall is presently Manager of Environmental Affairs & Safety for Agip Petroleum Co. Inc., with drilling and production operations in the Gulf of Mexico. Mr. Dougall's background includes consulting for Pilko & Associates, Inc., where he conducted environmental risk assessments for mergers and acquisitions, environmental compliance audits and permitting activities. His background also includes corporate and field assignments for Phillips Petroleum Company, including several years coordinating Phillips' permitting and compliance efforts in California. He received his Master of Environmental Science and B.S. degrees from the University of Oklahoma and is a registered professional engineer in Oklahoma and Arkansas.

Jill Barson Gilbert

Jill Barson Gilbert is a Technical Advisor at Pilko & Associates, Inc. Ms. Gilbert's environmental management experience includes positions in industry before joining Pilko & Associates, where she consults to industry, attorneys, and investment bankers on regulatory and environmental issues. Her background includes site investigations, remediation, waste management programs, environmental audits, environmental policies and procedures, and regulatory analysis. Ms. Gilbert's credentials also include the permitting of projects such as chemical plants, land disposal operations, hazardous waste facilities, and PCB incinerators.

Ms. Gilbert is active in the Air & Waste Management Association and is a member of the Faculty at the University of Houston—Clear Lake. She holds an M.S. in Environmental Management from the University of San Francisco and an A.B. from Miami University.

William P. Gulledge

William P. Gulledge is a Consultant in the Arlington, Virginia, office of Tillinghast, a Towers Perrin company. He specializes in environmental risk management and has over 15 years of experience in environmental management and pollution control activities.

Prior to joining Tillinghast, Mr. Gulledge was Vice President of Front Royal Insurance Company and Environmental Insurance Management. In this position, he was in charge of environmental insurance product development, underwriting criteria development, and claims management. Mr. Gulledge also served as the company's chief technical advisor for environmental analysis.

EIL products developed at Front Royal include coverage for petroleum storage tanks, golf courses and related recreational facilities, and lender liability. Mr. Gulledge also managed consulting services for Environmental Insurance Management. These services included preparing environmental liability site assessments, managing environmental claims (including identifying cost-effective and innovative cleanup technologies), and assisting states in establishing cleanup cost reimbursement programs for underground storage tank releases.

Mr. Gulledge serves in leadership positions on several environmental professional associations. He is President of the Institute for Environmental Auditing and is leading new activities to develop standards for auditing environmental management programs. With the American Society for Testing and Materials, Mr. Gulledge is leading efforts to develop cleanup procedures for old waste disposal sites.

Mr. Gulledge formerly was Associate Director of Environmental Programs for the Chemical Manufacturers Association (CMA). With CMA he developed policy and directed legislative and regulatory response for clean air issues, the environment, health and safety assessments, and superfund implementation. Mr. Gulledge also assisted members with programs to reduce toxic air emissions, manage and treat hazardous waste, and monitor for environmental compliance and pollution reduction effectiveness.

Prior to joining CMA, he served as a Senior Program Manager with the U.S. Department of Energy at the Pittsburgh Energy Technology Center. Mr. Gulledge developed control technology options for coal conversion and combustion processes and other alternative energy programs.

Mr. Gulledge holds a Master of Science degree in Environmental Sciences and Engineering from American University in Washington, D.C. He completed his undergraduate work in Political Science at American University.

A frequent participant in regional and national pollution liability and environmental management forums, Mr. Gulledge has prepared numerous publications on these issues. These publications address underground storage tank management, toxic air pollutant controls, environmental auditing, hazardous waste characterization and disposal, investment opportunities in environmental management, and environmental impairment liability and risk management.

Michael D. Henke

Michael D. Henke is a Senior Program Manager in the consulting firm of Pilko & Associates, Inc. He manages and conducts environmental audits and risk assessments in large and small facilities across the United States, for heavy and light manufacturing industries, hazardous waste disposal facilities, and so on. For several years Mr. Henke coordinated all of Pilko & Associates' risk assessment activity throughout the United States, Canada, and Europe.

Mr. Henke has broad experience in manufacturing, environmental affairs management, and environmental auditing. With Gulf Oil Chemicals (GOCHEM), he spent several years in plant operations and engineering. Subsequently, at GOCHEM headquarters, Mr. Henke instituted and conducted the company's environmental audit program, which covered 70 manufacturing facilities and laboratories throughout the United States.

Mr. Henke received his B.S. in Engineering Science from the University of Notre Dame, and his M.B.A. from the University of St. Thomas.

Douglas J. Hileman

Douglas J. Hileman is a Managing Associate with Pilko & Associates, Inc. in Los Angeles, California. His areas of expertise include clients' programs, site assessments, lender consulting, training programs, real estate, and litigation support. Mr. Hileman teaches environmental auditing in the certificate programs at the University of California Extension, Irvine. Doug is a member of the Steering Committee of the Environmental Auditing Forum, a Registered Professional Engineer, and a California Registered Environmental Assessor. He holds an M.B.A. from Cleveland State University and a B.S. in Chemical Engineering from Rose-Hulman Institute of Technology.

Lisa E. Lowe

Lisa E. Lowe is the Corporate Director of Environmental Regulatory Services at Sterling Winthrop Inc. (a subsidiary of Eastman Kodak). In this position, Ms. Lowe provides technical assistance to Sterling's manufacturing locations and is the project manager for several remediation projects. Ms. Lowe also is one of five team members responsible for conducting Environmental, Safety, and Health (ESH) assessments at Sterling's domestic and international locations, and is also a member of Eastman Kodak's ESH Assessment Team. She has more than 13 years experience in the hazardous waste, environmental auditing, risk assessment, and Superfund arenas. At Camp Dresser & McKee Inc. (CDM), Ms. Lowe was the Region II Assistant Manager for the REM II contract. She has managed audit programs, Phase I/II assessments, RI/FS projects, risk assessments, and site investigations, and has prepared audit training programs and several manuals for industry and government agencies. Ms. Lowe is a member of the American Society for Testing and Material subcommittee to prepare an "environmental due diligence" standard, and is also a member of the American Chemical Society, Society of Risk Assessment, and Air & Waste Management Association.

Linda M. Micale

Linda M. Micale is the principal of Micale Regulatory, a firm specializing in environmental regulatory programs. She has a Master of Sciences degree in Renewable Natural Resources from the University of Arizona. She has over

8 years of experience in assisting clients with complex issues involving environmental policy, law, and regulations. She specializes in regulatory analysis, planning, and compliance management and in identifying aggressive, cost-effective solutions to such issues. Her areas of expertise focus on RCRA, CERCLA, and EPCRA; however, Ms. Micale also provides analysis on waste minimization, pollution prevention, and hazardous communication requirements. Ms. Micale has managed over 100 projects driven by regulatory requirements in the western United States. She has served in the capacity of regulatory services manager in the Tuson office of Dames & Moore.

Ms. Micale authored a chapter on generation and disposal of hazardous waste and coauthored a chapter on medical wastes for *Hazardous Materials Toxicology: Clinical Principles of Environmental Health*, Sullivan and Kreiger, eds. William & Wilkens, 1991. She edited the 1985 publication, *Managing Water Quality in a Water-Scarce State*, for the Arizona Academy in Phoenix, Arizona, and previously coauthored a paper with Ms. Pamela Beilke for the 1988 Society of Mining Engineers Annual Meeting held in Phoenix, entitled, "The Impact of Arizona's Environmental Quality Act on the Mining Industry."

Sam N. Popowcer

Sam N. Popowcer, P.E., DEE, received his Bachelor of Science in Chemical Engineering from Illinois Institute of Technology. After serving 3 years in the U.S. Navy, he worked for Allied Chemical Corporation for more than 30 years. There he served in various line and staff functions, including Plant Manager, and Manager—Maintenance and Plant Engineering.

Mr. Popowcer recently retired from BCM Engineers, Inc. after 12 years, where he performed energy audits and industrial facility environmental assessments. He is a Registered Professional Engineer in Pennsylvania, a Registered Environmental Assessor in California, and Diplomate in the American Academy of Environmental Engineers.

Mr. Popowcer is now with Integrated Environmental Services principally involved in Environmental Regulatory Compliance, Documentation, and Training.

Frank J. Priznar

Frank J. Priznar is an internationally recognized authority on environmental auditing. Currently the Vice President and Manager of Roy F. Weston's Washington, D.C. office, Mr. Priznar is co-founder and past president of the Institute of Environmental Auditing. He has completed over 100 environmental audits worldwide for diverse commercial and government clients who seek ways to more effectively administer their environmental management responsibility. Mr. Priznar is a frequent speaker and writer on modern environmental management tools and techniques. He serves on the editorial boards of *Environmental Management*, published by Springer-Verlag, and *Environmental Testing and Analysis*, published by the Target Group. He was also editor of *Working Papers*, the only newsletter devoted to environmental auditing, from 1986 to 1992.

Mr. Priznar was formerly the Director of the Environmental Health and Safety Division at TechLaw, Inc. He created and led the Technical Environmental Compliance Health and Safety Practice at Booz, Allen & Hamilton, Inc. and was formerly the Director of the Analytical Services Division at Normandeau Associates, Inc.Frank J. Priznar

Paul D. Rosenstock

Paul D. Rosenstock is currently a prinicipal in Creative Regulatory Solutions, Inc., an environmental and industrial hygiene consulting company. Prior to entering consulting in 1990, he worked for 30 years in the chemical and pharmaceutical industries, of which 26 years were with Rohm and Haas Company, Inc., and Rohm and Haas Delaware Valley Inc. Most of his career has been in manufacturing where he served in various staff and line functions, including Environmental Manager, Regulatory Affairs Manager, Utilities Manager, and Technical Director. His areas of expertise are environmental auditing, permitting, waste reduction and minimization, process development, process safety, and project management. He has a B.S. in Chemistry from the Polytechnic Institute of Brooklyn and a Ph.D. in Chemistry from Pennsylvania State University. He is a Certified Industrial Hygienist in General Practice, a Certified Hazardous Materials Manager (Master Level) and a Registered Environmental Professional. Finally, he is active in numerous technical organizations and is Chairman Emeritus of the Water Resources Association of the Delaware River Basin.

Jacqueline Shields

Jacqueline Shields is currently an Assistant Professor at the University of Texas School of Public Health, MPH Program at San Antonio, where she teaches graduate courses in Environmental Health. Prior to coming to San Antonio, she taught at the University of Houston–Clear Lake in the Environmental Management Program and for 3 years in the Chemical Engineering Department and School of Pharmacy at the Universidad Autonoma de Guadalajara, in Guadalajara, Jalisco, Mexico. She has a B.A. in Chemistry from McMurry College, an M.S. in Civil Engineering from the University of Texas–El Paso, and a Ph.D. in Civil Engineering from the University of Oklahoma. Her research work and publications have included statistical analysis studies of ambient air levels of heavy metals, carbon monoxide, and ozone in selected cities in Texas. Her current research work is focused on ambient air quality in the Texas–Mexico border region and exposure assessments. In addition to her academic career, Dr. Shields worked for many years as a chemist for state and local government agencies and also in the private sector. Her work in the environmental field began in the 1960s and has continued uninterrupted to the present. She is a member of the Institute of Environmental Auditing, the Air & Waste Management Association, and several other environmental professional organizations.

Index

Acid deposition, baseline data, 96–98
Acid rain, 56
  acid precipitation monitoring network, 66
Action plans, 37–38
  documentation system in, 38
Administrative audit, air permit audits, 116, 118–122
Agency for Toxic Disease Registry, 95
Air and Waste Management Association, 4, 67
Air dispersion modeling, 69
Air emissions checklist, 33
Air emissions management, 199–202
  hazardous air pollutants, 201–202
  National Ambient Air Quality Standards, 199
  National Emission Standards for Hazardous Air Pollutants, 200
  New Source Performance Standards, 199–200
  odor control, 202
  permits, 202
Air monitoring data, 83–86
  assessment of monitoring program, 85
  issues related to, 85–86
  regional monitoring networks, 83–84
  review of air monitoring information, 86
  sources of data, 83
  use of data, 85
Air permit audits, 115–129
  administrative audit, 116, 118–122
  air permits under Clean Air Act Amendments, 125–129
  audit of permit requirements, 128
  basic requirements for, 115–116
  checklist for, 117
  compliance audit, 122–125
  regulated substances, 118–119
  types of permits required, 116
Air Pollution Control Association, 4
Air Pollution Engineering Manual, 67
Air quality
  air pollution audit protocol, 72–78
  air pollution questionnaire, 79–82
  air quality models, 67
  ambient air concentrations, 84
  criteria pollutants, 69, 84
  data for audit, 69–70
  data sources, 65–69
    Air Pollution Engineering Manual, 67
    air quality models, 67
    AP-42 Handbook, 67
    databases, 66–67
    environmental agency records, 67–68
    health effects data, 68–69
    health effects documents, 68
    information from states, 68
    private organizations, 68
    software for, 67
  fugitive emissions, 70
  indoor air data, 70
  non-criteria pollutants, 69–70
  particulate matter, 69
  secondary emissions, 70
Allied Chemical Corporation, 3–4
Ambient air concentrations, 84
Ambient air quality audit, case example, 263–265
American Petroleum Institute, 68
American Society for Testing and Materials, 15
Amoco, 4
AP-42 Handbook, 67, 93, 119
Asbestos, 179–180
  asbestos management at refineries/petrochemical facilities, 203–204
  regulatory requirements, 180
  types of building materials, 179
Asbestos Hazard Emergency Response Act, 158
Attorney-client privilege, and confidentiality, 29–30
Auditor
  and boundaries of auditing process, 57
  final selection of, 62
  hiring criteria, 57–61
    form for, 58

Auditor (*Continued*)
performance, 59–61
qualifications, 59
references, 61
hiring options, 55–56
Audit program, evaluation of, 38–39
Audit report, 36–37, 60, 143–144, 189–192
body of, 191–192
computer-generated report, 36, 37
environmental affairs management, 191
introduction of, 190–191
oral report, 37
summary of, 191
written reports, 37
Audit team, 28–29, 143, 187–188
environmental health audit, 239
external auditors, 28–29, 56
internal auditors, 28, 55
leader, role of, 44
mixed team, 56
training of, 187–188

Baselines
and Clean Air Act Amendments, 90–101
decision making about, 90
descriptive baselines, 87–88
difficult baselines to establish, 90
establishment of, 88–90
indirect baselines, 88
numerical baselines, 87, 89
preferred type of, 89
requirements
listing of requirements, 104–114
Title I, 92–94
Title II, 94
Title III, 94–96
Title IV, 96–98
Title V, 98
Title VI, 98–100
Title VII, 100–101
Benzene, 70
benzene service, 201
National Emissions Standards, 200–201
Biased ranking approach, 25–26
Biological treatment, 132
Biosludge, 132

California Bill, 7
Centers for Disease Control, 68, 95
Centralized audit programs, 50
Certified Industrial Hygienist, 59
Characteristic wastes, 154
Checklists, 43, 188
in self-audits, 50
summary checklists, 30
Chemical Manufacturers Association, 68, 135
CHEMTREC, 240
Chlorofluorocarbons
baseline data, 98–100
future actions, 84
and ozone depletion, 213, 216
Clean Air Act, 7, 10, 24, 199
Clean Air Act Amendments, 10, 11, 17, 30, 57, 65, 133–137
adjustments in auditing due to, 101–102
audit of permit requirements, 128
audit types of, 13
and baselines, 90–101
emissions audit, aspects of, 135–137
and environmental health audit, 220, 227–229
future view, 134
general audit form, 127
impact of, 134–135
major provisions of, 126–127
permitting program, 134
problems/constraints facing manufacturers, 134–135
and property transactions, 152–153
Titles of, 133
and use of external auditor, 56
Clean closures, 178
Clean Water Act, 7, 115, 197, 222
and property transactions, 153–154
Community
community exposures, 238
community health effects baseline, 90
community warning systems, 84
Emergency Response Plan, 121
perception of facility, 120–121
Community-right-to-know, 204–205
Compliance audit, 132–133
air permit audits, 122–125
Compliance environmental audits, 16
Comprehensive Environmental Response Compensation and Liability Act, 6, 10, 17, 195, 197, 225–226
and property transactions, 156–157
Computer-generated report, 36, 37
Confidentiality
attorney-client privilege, 29–30
in environmental audit, 29–30
work product doctrine, 29, 30
Control Technology Guidelines, 134
Council on Environmental Quality, 13
Criteria pollutants, 69, 84

Databases
access by auditor, 60

functions of, 66
health literature, 95, 240
on air emissions, 66–67
Data evaluation, in audit process, 32–36
Decentralized audit programs, 50
Department of Justice, 7
and environmental audits, 11
on prosecutions, 14–15
Descriptive baselines, 87–88
Desk audit, 213
Directory and Resource Book, 67
Disclosure, in environmental audit, 30
Document reviews, in quality assurance audit, 47–48
Documents, related to environmental audit, 48
Due diligence assessments, 159–168
field investigation, 163–164
objectives of, 159–160
and overcoming deal killers, 165–166
remediation, 164–165
remote sensing in, 160
risk assessment in, 160–163

Earth Day, 10
Electric Power Research Institute, 68
Electric steam generating units, 96
Emergency Planning and Community Right-to-Know Act, 204, 207
Emergency Response Plan, 121, 124
Emissions inventory program, 92, 136
Emission sources, types of, 123
Environmental assessment, use of term, 11, 186
Environmental audit, 137–144
areas addressed during, 27
checking of permits, 141
emissions, measurement of, 140
future view, 8–9
history of, 3–9
important influences on, 16–17
and management support, 22, 24
material balances in, 141
objectives of, 21–22
preacquisition audit, 186
quality assurance for, 41–53
reasons for not conducting audit, 138
reasons for use, 54–55
and recordkeeping, 139–140
types of audits, 16
use of term, 11, 186
Environmental auditing policy, 42
Environmental auditing policy statement, 6–7, 13–14, 226
aspects of, 14
development of, 13–14
Environmental Auditing Roundtable, 4, 43
Environmental audit process, 141–144
action plans, 37–38
audit report, 36–37, 143–144
audit team, 28–29, 143
confidentiality, 29–30
data evaluation, 32–36
data gathering programs, 142
disclosure, 30
evaluation of audit program, 38–39
follow-up, 144
phased approach, 142
planning audit, 24–28
preaudit activities, 31–33
protocol for audit, 30–31
remedial action responsibilities, 144
scheduling of audit, 28, 31, 143
site visit, 34–35
steps in, 23
Environmental Cleanup Responsibility Act, 157
Environmental Data Resources/Toxicheck, 67
Environmental health audit
audit team, 239
conducting audit, 237–241
data collection, 239–240
legislation related to, 220–229
structure of audit, 229, 233–237
consequences section, 235–236
documentation section, 237
ecosystem section, 233–234
exposure section, 234–235
population section, 229, 233
social section, 236–237
treatment section, 235
Environmental impact reports, scope of, 170
Environmental impairment liability insurance assessments, 16
Environmental liability site assessments, 6
positive aspects of, 6
Environmental management program audits, 16
Environmental Protection Agency
and beginning of environmental auditing, 4
creation of, 10
environmental auditing policy statement, 6–7, 13–14
Environmental Responsibility and Clean-up laws, 138
Environmental risk
operational risk, 151
personal injury risk, 151
public policy risk, 151

Environmental risk (*continued*)
  residual risk, 150–151
Environmental Risk Imaging and Information Center, 67
External reviews, for quality assurance, 46

"Factors in Decisions on Criminal Prosecutions for Environmental Violations in the Context of Significant Voluntary Compliance of Disclosure Efforts by the Violator," 8, 14, 30
Federal Insecticide, Fungicide and Rodenticide Act, 204
  environmental review aspects, 204
  and property transactions, 158–159
Federal Pollution Prevention Act, 137, 207, 208
Federal Water Pollution Control Act Amendments, 10, 17, 196
Field audit, for ozone-depleting chemicals, 214–215
Follow-up, in audit program, 52, 144
Fourier transform infrared spectroscopy, 135
Freedom of Information Act, 68
Freon, 245
Fugitive emissions, 70, 140
Fundamentals of Environmental Compliance Inspections, 36
Fungicide, legislation related to, 158–159

General Motors, 4
Generator Identification Number, 179
Grandfathered units, 202
Green audits, 16
  diseases caused by, 179–180
Groundwater, and demolition/razing, 178–179
Groundwater contamination, 194–195
  regulation of, 194–195

Hazardous air pollutants, 201–202, 240
  regulations, 56, 57, 94–95, 119
    baseline data, 95–96
Hazardous and Solid Waste Amendments, 155, 198, 207
  and property transactions, 154–155
Hazardous waste, 154–156
  characteristics wastes, criteria for, 154
  clean closures, 178
  fines for, 155
  generators, nature of, 155
  legislation related to, 154, 155
  listed wastes, 154
  types of substances, 193
  underground storage tanks, 155
  waste treatment units, 177–178
Health audits. *See* Environmental health audits
Health effects data, 68–69
Health effects documents, 68
Health records, of workers, 120
Hickman v. Taylor, 30
Hydrochlorofluorocarbons, 215
Hydrofluorocarbons, 215

Indoor air data, 70
Innocent landowner clause, 168
Insecticides, legislation related to, 158–159
Institute for Environmental Auditing, 7, 13, 15
Internal reviews, for quality assurance, 45–46
International Chamber of Commerce, 7, 42
International situations, and auditor, 59
Interviews, in quality assurance audit, 47

Lake Erie, 153
Landfills, 177–178
  closures for, 178
Legislation, 27
  Clean Air Act, 7, 10
  Clean Air Act Amendments, 10, 11
  Clean Water Act, 7
  Comprehensive Environmental Response Compensation and Liability Act, 10
  Federal Water Pollution Control Act Amendments, 10
  future view, 17
  goal of current laws, 132
  impact of, 15–17
  most significant legislation, 11
  National Environmental Policy Act, 10
  performance-based regulations, 12–13
  regulatory wheel, 131
  Resource Conservation and Recovery Act, 10
  Safe Drinking Water Act Amendments, 10
  Superfund Amendments Reauthorization Act, 10
  technology-based regulations, 11–12
  Toxic Substances Control Act, 10
Lender liability, 166–168
  project requirements, 167

theories of, 166–167
Liability provisions, Superfund, 156
Listed wastes, 154
Love Canal, 154
Lowest achievable emissions rate, 134

Major sources, meaning of, 133–134
Management support, and environmental audit, 22, 24
Material balances, 141
Material safety data sheets, 205, 214, 247, 251
Maximum achievable control technology, 134, 211–212
Maximum contaminant level goals, 195
Maximum contaminant levels, 195
MEDLINE, 95, 240
Methane, 100
Mini-health risk assessment, 136
Montreal Protocol, 98, 213
Multi-media Compliance Audit Inspection Procedure, 142

National Aeronautics and Space Administration, 98
National Ambient Air Quality Standards, 12, 65, 92, 118, 152, 199, 240
National Emissions Data System, 66
National Emission Standards for Hazardous Air Pollutants, 118, 152, 200–201, 240
National Environmental Policy Act, 10, 170, 227
National Oceanic and Atmospheric Administration, 98
National Pollutant Discharge Elimination System, 13, 134, 196
National Response Center, 197, 240
New Source Performance Standards, 199–200
New Source Performance Standards permit, 67, 68
Non-criteria pollutants, 69–70
Notices of violation, 119
Nuclear generating stations, volatile organic chemical inventory, case example, 251
Numerical baselines, 87, 89

Occidental Petroleum, 4
Occupational Health and Safety Administration, 17, 68–69
Odors
at facility, 123
regulation of, 202
Off-site cleanup, 195
Oil spills, legislation related to, 153
Olin, 4
Operational risk, meaning of, 151
Oral report, 37
Ozone, 66
concentration data, 84
ozone depleting chemicals, 213
Ozone-depleting chemical audit
case example, 245–246
field audit, 214–215
preaudit activities, 214
report preparation, 215–216
source reduction recommendations, 216

Particulate matter, 69
Peer review, of audit report, 44
Penalties
baseline data, 100
fines for hazardous waste, 155
notices of violations, 119
Performance-based regulations, 12–13
Personal injury risk, meaning of, 151
Pesticides
reporting/labeling requirements, 204
and water pollution, 154
Phased chemical usage study, case example, 247–248
Planning audit, 24–28
audit scope, defining, 24–28
frequency of audit, 28
scheduling audit, 28
Point sources, 153
Pollution prevention
barriers to, 145
meaning of, 144–145
options in, 147
requirements for, 146
and states, 145
and waste minimization, 144–148
Pollution Prevention Benefits Manual, 144
Pollution Prevention Plan, 137
example of, 210–211
Pollution Prevention Resource Manual, 139
Polychlorinated biphenyls, 180–181
audit team review, 203
refineries/petrochemical facilities guidelines, 202–203
regulatory aspects, 181, 202–203
sources of, 180–181
Potentially responsible parties, 156
Private sources, for air quality data, 68
Professional Engineer, 59

Property life cycle, 168–175
  demolishing of site disposition, 174–175
    environmental impacts, 174–175
  development of raw land, 169–171
    environmental impacts, 169–170
    project implementation, 170–171
  expansion/modification of existing facility, 171–174
    environmental impacts, 171–173
    project implementation, 173–174
  site assessments, 172, 173
Property transactions
  and Clean Air Act Amendments, 152–153
  and Clean Water Act, 153–154
  and Comprehensive Environmental Response Compensation and Liability Act, 156–157
  due diligence assessments, 159–168
  environmental issues related to, 175–183
    active/inactive waste sites, 177–178
    asbestos-containing materials, 179–180
    polychlorinated biphenyls, 180–181
    residual materials and wastes, 181–182
    soil/groundwater contamination, 178–179
    underground storage tanks, 176–177
  environmental risks in, 150–151
  and Federal Insecticide, Fungicide and Rodenticide Act, 158–159
  fertilizer plant purchase, case example, 260–262
  and Hazardous and Solid Waste Amendments, 154–155
  innocent landowner provision, 168
  lender liability, 166–168
  property life cycle, 168–175
  property transaction audits, 16
  and Resource Conservation and Recovery Act, 154–155
  and states, 157–158
  and Superfund Amendments and Reauthorization Act, 156–157
  and Toxic Substances Control Act, 158
Protocol for audit, 30–31
Publicly owned treatment works, 141
  publicly owned treatment works emissions, 96
Public policy risk, meaning of, 151

Quality assurance audit, 41–53
  and components of audit process
    documentation of findings, 51–52
    follow-up, 52
    methodology of audit, 50
    objectives of audit, 49
    organizational position of audit program, 51
    performance measures, 52–53
    scope of audit, 49
    staffing of audit program, 50–51
  design/management of audit process, 43–45
  document reviews, 47–48
  external reviews, 46
  internal QA reviews, 45–46
  interviews in, 47
  objectives of QA audit, 47
  observation of audits, 48
  organizations endorsing quality assurance, 42–43
Questionnaires, 31

Radon, legislation related to, 158
Ranking systems, 25–26
  biased approach, 25–26
  unbiased approach, 26
Reasonably available control technology, 134
Recordkeeping
  and environmental audit, 139–140
  importance of, 139, 140
  requirements related to, 140
References, of auditor, 61
Refineries/petrochemical facilities
  air emissions management, 199–202
  asbestos management, 203–204
  community-right-to-know, 204–205
  equipment found in, 185
  on-site waste management, 193–195
  pesticide management, 204
  polychlorinated biphenyl guidelines, 202–203
  spill containment, 197–198
  underground storage tanks, 198–199
  water/wastewater management, 195–197
Refineries/petrochemical facilities environmental audit
  audit report, 189–192
  audit team, 187–188
  groundwater contamination, 194–195
  information gathering, 188–189
  objectives of, 186–187
  scheduling of audit, 188
  site review, 189
Refinery environmental audit, case examples, 254–259
Refrigerants, 215
Regional monitoring networks, 83–84
Release, definition of, 205

Remediation, in due diligence assessments, 164–165
Remote sensing, in due diligence assessments, 160
Report. *See* Audit report
Residual materials and wastes, 181–182
Residual risk, meaning of, 150–151
Residual wastes, 181–183
  disposal of, 182–183
  types of, 181
Resource Conservation and Recovery Act, 6, 10, 17, 24, 25, 115, 154, 207, 226
  and property transactions, 154–155
Risk assessment, aspects of, 219–220
Rodenticide, legislation related to, 158–159

Safe Drinking Water Act, 195
Safe Drinking Water Act Amendments, 10
Scheduling of audit, 28, 31, 143, 188
Secondary emissions, 70
Security and Exchange Commission, enforcement actions of, 3–4
Self-audits, aspects of, 50
Site review
  aspects of, 189
  during audit, 34–35
  observations during site review, 190
Software, air emissions information, 67
Soil contamination, 178–179, 194–195
  legislation related to, 194
  refineries/petrochemical facilities, 194–195
South Coast Air Quality Management District, 153, 215
Spill containment
  refineries/petrochemical facilities, 197–198
  review of facility, 197–198
Standard Industrial Classification Codes, 157
Standards, private, development of, 15
States
  air quality data, 68
  audit mandates, 138
  pollution prevention activities, 145
  and property transactions, 157–158
Stormwater, 207–208
  and water pollution, 153–154
Sulfur dioxide, 66
Superfund Amendments and Reauthorization Act, 6, 10, 17, 68
  development of, 156
  innocent landowner clause, 168
  liability provisions, 156
  and property transactions, 156–157
  Superfund sites, 157
  trends related to, 157
Superlien, 157
Supervision of audit, 43–44
Synthetic Organic Chemical Manufacturing Industry, 119

Technology-based regulations, 11–12
Threshold planning quantity, 204–205
TOMES, 95
Toxic Chemical Source Reduction and Recycling Report, 209–210
  contents of, 209
Toxic Constituent Leaching Procedure, 140–141, 154
Toxic Release Inventory System, 66
Toxic Substances Control Act, 10, 17, 202, 227
  and property transactions, 158
Traffic inventories, 92–93
Treatment, storage, and disposal of hazardous waste, 115, 155
Trichloroethane, 245

Unbiased ranking approach, 26
Underground storage tanks, 176–177
  considerations in inspection of, 177
  refineries/petrochemical facilities, 198–199
  regulation of, 155
  removal of, 176–177
Utility unit, 97

Vegetation, on/near site, 124
VISTA/Environmental Audit, 67
Volatile organic compound emissions, 93, 94, 115, 118
  case example, 251

Waste, definition of, 132
Waste management, 193–195
  off-site cleanup activities, 195
  on-site facilities, 193–194
Waste minimization, 144–148
  case examples, 249–250, 252–253
  legislation related to, 145
  minimization options, types of, 146
  and pollution prevention, 144–145
  and reduction of waste generation, 146
  waste minimization audits, 16
Waste Minimization and Control Act, 145

Waste sites, 177–178
Waste streams, analysis of, 140–141
Wastewater management, 195–197
  audit team review, 197
  management issues, 196–197
Water pollution
  pesticide contamination, 154
  point sources, 153
  and stormwater, 153–154
Water Quality Act, 196
Work product doctrine, and confidentiality, 29, 30
Written reports, 37, 46
  aspects of, 51–52